D1348400

CPM Construction Scheduler's Manual

Chapter 9 Contract Law **113**
Legal Aspects of the Project Scheduler 113
Outside Delays 115
Legal Notices 118
Notice of Completion 118
Notice of Cessation 118
Notice of Non-Responsibility 118
Preliminary Notice 119
Mechanics Lien 120
Notice to Owner 122
Stop Notice 123

Chapter 10 Construction Law **125**
Project Schedule Labor Use 125
Labor Codes 125
OSHA 129
Public Works 131
Lien Period Timetables 132

Chapter 11 Schedule Estimating **135**
Takeoffs 135
Quantity Survey 136
Long-Lead Items 137
Cost Tracking 138
Audit Trail 139

Chapter 12 Contract Conditions **143**
General Conditions 143
Contract Specifications 144
Shop Drawings 147
Long-Lead Items Purchase Orders 149

Chapter 13 Preconstruction Planning **151**
Progress Schedule Regimentation 151
Bid Award Prior To Commencement 152
Schedule Regimentation 153
Shop Drawings Log 154
Transmittals 154
Schedule Planning 155

Chapter 14 Project Operations **157**
Recycling The Schedule 157
Cost Monitoring 158
Schedule Operations Analysis 159
Daily Field Reports 160
Change Orders 163
Cost-loaded CPM 163
Sorts 164
As-Planned, As-Builts, and Adjusted Schedules 166

Chapter 15 Software Program **167**
Installation 167
Data Entry 167
Entering Numbers 170
Entering Formulas 172
Logic Formulas 173
Using String Operators 175
Menu Commands 177
Saving Your Work 178
Procedures 180
Changing Data 180
Database Terminology 182
Using Databases 183
Building a Database 184
Sorting Databases 186
Sorting With Key Sorts 186
Double Keys Sorts 187
Extra Keys Sorts 187
Datatables 188
Datatable Range 188
Using What-If 189
Directories / Sorts Report 190
Sort By Activities 190
Sort By Events 195
Sort By I-J Numbers 200
Sort By Job Logic 205
Sort By Total Float / Late Start 210
Cost By Activity Number 215
Schedule Of Anticipated Earnings 219
Sort By Early Starts 223
Daily Field Reports 227
Bar Chart By Early Sort 229
Network Timeline 234
Shop Drawings Log 236

Submittal Items Tracking 238
Correspondence Transmittals 240
Change Order Tracking 242
Audit Trail Tracking 244
Contract/P.O. Summary 246
Saving Your Sorts 248
Canceling A Command 249
Closing A Sort 249
Opening A Sort 250
Exiting ***FastPro*** and **Lotus** or **Quattro** 250
Deleting Cells, Rows, & Columns 251
Merging & Splitting Cells 252
Merging Column Width 253
Merging Row Height 254
Changing Cell Alignment 254
Printing 255
Using ***FastPro*** in **Windows** 255
Opening A Window 256
Copying & Moving Between Windows 256
Moving Windows 257
Closing Windows 258
Maintaining Schedule Sorts 258
Saving Sorts To A Different Subdirectory 260
Saving Sorts To A Different Disk Drive 260

Chapter 16 Project Scheduler Consultant 263
Running The Shop 263
Tax Benefits 265
Home Office Deduction 267
Building A Clientbase 268
Business Management 273
Accounting 276
The Bookkeeping Process 277
Your Financial Statements 280
New Technology 286

Chapter 17 Scheduling Contingencies 289
Scheduling Contingencies 289
Scheduling Changes 290
Constructive Changes 291
Changed Conditions 291
Schedule Acceleration 292
Project Schedule Productivity Losses 295

Suspension of the Project Schedule 296
Change Order Pricing 297
Project Schedule Conflicts 297

Chapter 18 Conclusion **305**
Post Script 305
Adios 307

Appendix A **311**
Project Production Abbreviations

Appendix B **325**
Industry Associations

Appendix C **331**
EJCDC Table of Contract Standards

Index **345**

Acknowledgments

I would like to take this brief moment of opportunity as our trails cross, to thank the other professionals in various businesses and construction industry experts who had a hand in bringing this book and integrated software program into existence. Each of them are good and caring people whom it has been my great privilege to have known and worked with. Thank you all for your information, support, and help. You helped make this book a reality.

Robert Davison, Emerald Cut Graphics, Oak Harbor, Washington
Gary Griffith, Computer Division Director, Cabrillo College, Aptos
John DeCourcey, Senior Building Inspector, Santa Cruz County
Fred Sibley, President, Power Engineering Inc., Los Angeles
David Burton, Construction Management Director, Cabrillo College, Aptos
John Alexander, Vice President, Dean Witter Reynolds Inc., Santa Cruz
Roger Ludlow, RCI Certified Building Inspector, Santa Cruz
Daniel Ehrler, CEO, Santa Cruz Chamber of Commerce, Santa Cruz
David Ambrosini, Business Division Director, Cabrillo College, Aptos
Elza Minor, SBA Branch Office Director, Aptos
Chad Simmons, Director, Construction Management Institute, Soquel
Robert G Cados, Director, Construction Management Institute, Soquel

Finally, I would like to thank April Nolan, trade editor, McGraw-Hill Inc., without whose patience, help, support, and expertise the combined experience and knowledge of myself and all these good people would have never made the journey to your eyes, your mind, and your career.

As your career progresses, always remember: The buildings you help bring into existence will be your legacy far beyond your biological lifespan and decades after. Attention to fire, life, and safety protection for the public (the credo of the building inspector trade) should always precede the age-old drive for profits. If you adopt this mindset for your basic foundation of career ambition, good projects will come to you. Life is a mirror. You get out of it what you put into it. All that really matters when they are shoveling dirt on your coffin are:

The people who grieve at your passing;
The amount of good you left in your wake;
The buildings that were borne of your work.

Foreword

Ten years ago, when I first met Jonathan Hutchings, we were both working as site development engineers on a large semiconductor company's expansion project here in Silicon Valley. His secret of success in project scheduling was his skill in tight CPM velocity network diagramming that allowed for multiple critical path options. When the inevitable problems in activities production occurred, his network scheduling always seemed to have an operable plan B ready to go, with another plan C right behind it. Clients always seemed to like the way his project management skills would "bulletproof" their money interests.

I've learned a lot from him over the years, about the intricacies of multiple-path CPM project management, and he's constantly been harping at me to upgrade my company's computer systems. I know he's right about taking advantage of new technologies in construction management. Consistently, I see corporate owners pleased with his computerized cost-control systems that kept the project team on their toes regarding budget overrun restraints. Times have changed, and this industry has changed too. Nowadays it seems change is a daily part of my business. For instance, my new pickup has a factory-installed module for PC, fax, modem, and cellphone. It gives me a complete telecommunication office in the cab of my vehicle. I can enter field reports from my pickup at the job site, directly into the main scheduling program running on the war room computer miles away, via the modem and cellphone. I can remember a few years back, when Jon was telling me these things were coming, and now they're here. Moreover, these new technologies are in use everywhere in modern construction management, and one must master them to be competitive.

Over the years as our friendship has grown, I've watched Jon grow and evolve beyond the old construction management techniques to helping usher in new CPM technologies appearing in the industry almost daily. In addition to his management-consultant business, he became a state-certified teacher of business management, construction management, and microcomputer applications. His classes at the state university are always full, due in no small part, I'm sure, to his confidence and skill in making the fundamentals of construction management clear and easily understood by all. Jonathan is a positive, aggressive ex-Marine, and his degree of professionalism reflects it. His scheduling software program, ***FastPro CPM Scheduler***, is used in college level coursework. My company uses it daily, and I often find it more suitable for some of my projects than the higher-priced programs we use for multi-project scheduling. ***FastPro CPM Scheduler*** contains features found in programs costing much more, and handles a variety of my project scheduling tasks from residential to commercial work. As I understand it, a student version is to accompany this edition, and my professional advice to you is learn it, master it, use it. I say this because you, the reader, interested enough to pick up this book and check it out, must realize that the entire construction industry has changed drastically from the old days and the old ways, just as all businesses must, to survive the new age upon us. This book will help you learn how to make the new ways of change work for you. What you will learn from this reference manual today is real-world application stuff that you can use on the job tomorrow.

Everyone these days (unless they're living in a cave on the moon) knows that, to improve their business and remain competitive, they must learn to apply their skills by using the new technologies rushing at us at breakneck speed. In commercial construction projects today, I typically get blueprint plans running to 200 or more pages. Without computerized CPM, my business would fail within the quarter. Nowadays, project managers like Jon have computers running computers in the project office war room. Entire walls are covered in network diagrams, and multiple telephones are ringing incessantly. Estimators wear pilot-type headsets and walk around the room, plugging into modular phone jacks at different computer workstations, running the details of multiple jobs simultaneously. The amount of energy present in today's project war room makes the air feel charged with electricity. Decisions relating to expenditures of thousands of dollars are made instantly, every hour of the day. The sofa at the end wall is not for sitting on, it's for cat-napping on when the action starts going round the clock as completion date nears. Everywhere you look, terminals display computerized CPM programs like this one running at breakneck speed. Fast-tracking isn't just a type of scheduling, it's a way of life when projects turn profitable.

In project management, time is everything. Time is profit. Time is the difference between your company's being next year's income leader or being this year's history. CPM is the way we professionals manage time. And one of the best pros I know, at CPM scheduling, is the author of this book. This edition will definitely be on my shelf and this software is already a working part of my business. And if you're serious about carving out a successful career in construction management for yourself, my advice to you is to grab this very special book in a hurry, before they sell out.

Chad Simmons

CA Lic 444986
President, Simmons Engineering Company
Santa Clara, California

Introduction

Project scheduling is one of the hot new career jobs now growing within the modern, constantly updating construction industry. While lenders, owners, contractors, developers, project teams, inspectors, designers, engineers, suppliers, and a host of others each play an integral part in the construction process, the project scheduler is instrumental in making the production schedule either profitable or a financial disaster. As a licensed and certified construction management professional for over a decade now, I see this business from the inside out. Currently, almost no reference books are available that are specifically dedicated to CPM project scheduling, and there is almost no dedicated CPM computer software that is affordable to the student and small business owner. The CPM software currently on the market is geared for commercial to industrial work and averages about $4000 for a fully integrated programs. So McGraw-Hill hired me to fill the gap between what's available commercially and what's affordable.

Because the construction industry is so competitive as it is, one must be aggressive to be successful. And as my editor will tell you, I have all the social graces of a hand grenade and I'm not bashful. But even if I reincarnated Harry Houdini, and he and I started a magic show called Houdini & Hutchings, we wouldn't have the magic needed to stuff a four-hundred-dollar software program into a sixty-dollar reference manual. So for me to make the dollars work for my software programmer and my publisher (and still be affordable for you), we developed a working student version of CMI's commercial program ***FastPro CPM Scheduler***. The credit here goes to the software programmers at Construction Management Institute, Santa Cruz, California, who managed to scale down the essentials and integrate the program for this student version. They did a great job and you are definitely getting your money's worth.

This book/software package is unique in that it addresses the project scheduling field as it has never been addressed before. The software included in the book contains the essential sorts (summary reports) of a CPM velocity network schedule, which are integrated to provide a working network schedule incorporating program logic. You enter the data, the computer does the work. Use of this type of software program, and repeated exposure to this reference manual, will make the reader more efficient in the career field and more marketable as a professional CPM project scheduler. The skills learned from this book can be applied by any scheduler of production-type project execution. Current professional schedulers will find tons of bulletproofing available in the business combat, preconstruction planning and contract law chapters.

Students will find a two-semester college course condensed into this bookware product, which can be assimilated at their own pace. Small business owners, both in construction and production businesses, will find a $500 CPM software program in their hot little hands for $60. You will find, if you research, that this is currently the premiere book nationwide on CPM scheduling. There are others, but none as comprehensive, and that includes an integrated software program. This is it, the best CPM reference manual you will find anywhere.

While I'm sounding like a blowhard braggart here, let me explain why you are fortunate to have had this book come into your possession. In today's fiercely competitive business world, there just isn't room any longer for mediocrity. Being experienced and able to do the job just isn't enough anymore. Costs of materials and labor must be bid tight, and there's not much margin there to work with. So *time* is the only factor left to manipulate for producing profit. Accordingly, today's profits in construction projects are being made by superior CPM project schedulers, using modern systems such as fast-tracking. Even the best project planning and management cannot overcome the lack of a skillful CPM network project scheduler. In modern residential, commercial, industrial, and public works construction, it is the project scheduler who is *pivotal* in bringing the project into existence, on time and on budget.

The construction industry has changed in the last decade dramatically from companies who once dominated their markets by size, bidding, and bonding power to the modern, lean, downsized companies of today who use strategic personnel in definitive career positions for operations management, and contract all the field work out to licensed subcontractors. In the trades, these contractors are called *paper contractors*, because all they handle is paper. We also call them smart. No tools, no trucks, no employees. Their success formula is simple: Tight bid + lower overhead = we get the job. We make the profit through CPM fast-track scheduling. These types of paper contractor companies are currently enjoying a large growth in the construction industry because of their efficiency in CPM construction management. These are also the primary companies you want to target your project scheduler resume to.

In addition to new sophistication in modern construction management, the new age of technology has also brought in new materials, new technology, and powerful computer systems using dedicated software programs, opening the way for these and other specialists to pursue highly profitable, specialized careers in widely varying fields of construction management, both today and on into the coming century. You will see as we proceed that the career options for you to make a very healthy paycheck are expanding even as we speak.

I have earned my living in the construction industry for more than 20 years now, from laborer to apprentice, to journeyman, to foreman, to contractor; in management roles from estimator to project manager, to project scheduler, to state-certified instructor in construction management, business management, and microcomputer applications. In those two decades I have seen many changes that foreshadow even bigger changes in the future as the new century accelerates towards us. Of all the things we build with today, seventy-five percent have been invented within the last fifty years.

Of those, eighty percent have been invented in the last ten years. It is estimated that, of all the things we will be building with by the year 2005, ninety percent haven't even been invented yet. Graphite composite plastics that are stronger than steel, air-entrained lightweight concrete, and ultralight metal alloys are all space-age technology being used in modern construction today. And even as you read this, the clock on their obsolescence is ticking. *Paradigm* is a word that means the currently accepted parameters of any specific thing, such as an industry. And the construction industry's paradigm is changing as fast as the clouds above your head. More important, as Alan Toffler so correctly stated in his prescient book, *Future Shock*, the *rate* of that change is accelerating exponentially.

Worldwide, in residential construction, American Western Platform Framing is the accepted standard of excellence in homebuilding. After serving as an estimator and consultant for the insurance industry's reconstruction efforts in both the 7.6 Loma Prieta Earthquake of 1989, and the 7.8 Northridge-Los Angeles Earthquake of 1993, and having witnessed the devastation firsthand, I now understand why the rest of the world sends their building officials and industry representatives to America to learn our methods of construction. When one stops and reflects, the reason the world looks to us for leadership in construction is quite clear.

We build the best. That is not just my opinion, that is documented fact. The statistics: 61 casualties in Northridge-Los Angeles compared to 5,000 in Kobe, Japan. Both earthquakes rated 7.5 on the Richter scale. The difference is due to our highly evolved standards of construction management. In commercial and industrial construction, our methods of reinforced concrete and steel construction are unequaled in history. This is the last proud American industry still retaining global leadership, and justifiably so. May your career as a construction management professional in this proud industry bring you as much pride and personal satisfaction as mine has.

Remember, in matters of real-world applications of construction law, always research your data thoroughly, make sure it is as updated as possible, and consult an attorney. Further, please be informed that the subjects, theories and opinions contained herein are for general information purposes only and are not to be construed as legal advice. Both the publisher and the author assume no liability whatsoever, either expressed or implied, for business applications or any other usage of this instructional reference manual and software program.

Respectfully submitted,

Jonathan F. Hutchings, BCM
CA Lic 444806

01 July 1995

Chapter

1

Project Scheduler Overview

Work Parameters

Humankind's oldest and most noble profession is the construction of permanent shelter for human habitation and work or group assembly. Since the industrial revolution, construction projects of importance, size, and high cost have been organized and brought into existence by project teams working by a project schedule. Whether you are a newcomer entering a career as a construction management professional, a student pursuing a higher degree in construction management, or a seasoned project scheduler who needs a comprehensive reference manual with accompanying CPM computer software, this book/software product was produced to provide you with all the modern *nomenclature* (i.e.: how things work) of the fundamentals in CPM project scheduling and what it takes to complete a modern project on time and on budget, in a profitable manner.

The parties to any construction contract may be of either gender and may be singular or plural, such as individuals, partnerships, or corporations. However, traditionally, the American Institute of Architects uses singular, masculine pronouns as the accepted standard in text. Because this is the accepted legal industry standard, the author also uses this text format, but wishes to make absolutely clear to the reader that, although women traditionally have made up only a small fraction of construction management professionals, this fact has been changing in recent years. It is one of the author's personal and professional goals to provide career training for women in nontraditional career roles within construction management. The author and publisher both take pride in producing this and other reference books that provide comprehensive career training for all individuals interested in advancing in construction management careers, regardless of gender.

America being as big and diverse as it is, and this publisher being a premier marketing leader worldwide, brings me to a point before we begin where some area-specific terminology needs to be clarified. On the west coast, the construction project team member that oversees production is called the Project Manager (PM). On the east coast, the construction project team member that oversees the construction production is called the Construction Manager (CM). On the west coast, the construction project team member that develops and oversees the project schedule is called the Project Scheduler (PS). On the east coast, that project team member is called the Production Coordinator (PC). On the east coast they it wallboard, on the west coast we call it drywall. The list goes on and on. I believe that we construction management professionals do this just to confuse people starting out in the business, and to keep all the work to ourselves. This reference book is configured in west coast terminology and is based on the modern fundamentals of Western Platform Framing.

Before we define the modern project scheduler and that management position within the construction industry, it is necessary to explain the career's evolution and get a sense of how it came into existence, because of needs unfulfilled by traditional construction management techniques. Professional project scheduling evolved because of the inherent time-management inefficiencies in the traditional construction process of the fifties and sixties, which culminated in the late sixties with an national industrywide crisis in which most projects, both public and private, were running way over budget and long past their scheduled completion dates. The second biggest government building ever constructed, the Federal Triangle Project in Washington, D.C., will be 400,000 square feet when done, second in size only to the Pentagon. It is, at the moment of this writing, four years behind schedule and has doubled its original cost. It also serves as another example of this national industry-wide crisis of most capital projects running way over budget and long past their scheduled completion date. This problem has become so widespread that funders of capital projects in the mid-seventies began to demand that cost-effective methods be instituted in new types of construction management techniques before more capital projects would be funded in the future. At the time, many contractors and some architects were not trained in CPM and were unable to use efficient new CPM management skills, just at the time when the industry's changing paradigm demanded nothing less than skillful use of those new CPM skills. The need for a scheduling specialist surfaced.

The project scheduler career further evolved to counter the fact that architects and design consultants have built a reputation over the last two decades for not focusing accurately enough on the projected costs of labor and materials. Specifically, they have often been guilty of underestimating the project's production management costs to the project developer. If you are an architect and that sentence angers you, tough. You guys brought it on yourselves and you'd better wake up to the total costs involved for your clients, which means analyzing everything connected to your designs that's going to cost your client money to produce. Owners and their representatives that make up the project team were understandably concerned about this historical tendency of architects to take as little responsibility as possible for developing cost-tracking audit trails, producing network schedules, monitoring the project's subcontractor payments, or developing the project's quality control plan. All these project production control things, that make or break a project after it has left the designer's hands, have been traditionally left up to the owner by the design professionals. Something had to give, and owners began looking to hire those trained in CPM project scheduling for cost protection. Presto... the job of scheduling specialist appeared.

That said, in all fairness, I also need to point a finger at the other professionals involved in the project production: the building and specialty contractors. Activities subcontractors also have a long track record of operational deficiencies in their part of the traditional construction management process. Some contractors lacked up-to-date skills in modern computerized construction techniques, or the ability to understand and work with CPM network schedules. Some contractors understood the basics of CPM network scheduling but didn't think the new ways were important enough to invest money in and update their businesses. Eventually they were operating old-fashioned business infrastructures that couldn't handle the increasingly more complex and detailed CPM network schedules coming down the pike every day. Building contractors simply didn't keep up with change. If you're a building contractor and that last sentence angers you, tough. I'm a licensed building contractor too, and I know for a fact we're just as guilty as the design professionals in not looking after the client's needs as closely as we do our own. A lot of us builders have the attitude that we know it all and are resistant to change, eyeing it suspiciously as it keeps creeping closer.

This head-in-the-sand attitude by contractors has also cost owners money in the past, and has brought forth still another need for a professional project scheduler to counter the lack of new technology usage by contractors. Something had to give, and owners began looking to hire those trained in CPM project scheduling for protection from extra work orders and delay claims. Everyone loses when the project winds up in litigation except the lawyers. Now I've probably ticked off the lawyers too...

Additionally, in the past decade, as this accelerated rate of change swept through the construction industry, the jobs of project managers became more and more generalized. They were swamped between 14-hour days of managing the daily production and, in addition, interfacing with all the parties involved in the project. Suddenly there were simply not enough hours in the day for one person to physically handle all these duties and develop, administer, and track the project schedule as well. Hiring professional project schedulers and delegating the work load of developing and administering CPM project schedules was the logical outcome. Now, with a firm lock on the future, the professional project scheduler career is in full bloom. It is a sunrise career opportunity on the threshold of a bright and profitable future. We will always continue to build, and the world will always need professionals to manage that industry.

Even with today's sophisticated computer technology, there are no shortcuts for deleting a human project scheduler. If it were easy, everybody would be doing it and it wouldn't pay so well. Although it takes a conceptual mind trained in critical thinking, and an experienced hand in the ways of the construction industry, some *standard operating procedures* (SOP) will help you prepare for this job. These procedures will teach you to build your data into logic systems, sequentially interacting with each other, and by using dedicated project scheduling software, to organize your critical thinking to identify potential production bottlenecks. Following established scheduling SOP teaches you to make use of the workarounds, alternatives, options, constraints, and accelerators in CPM management as the *time manipulation* tools of the professional project scheduler. In today's fiercely competitive business world, profit and longevity come only to those who master the art of scheduling with flexible time management.

Because the current environment is so competitive, many companies have seen their profitability decline. But some companies continue to record superior rates of growth. These successful firms have recognized that work environments and the nature of the business have changed, and they have moved to accommodate these changes. A hard lesson was learned by the construction industry as a result of the recessions of the 1970s and early 1980s: increases in the cost and complexity of doing business dictate that all costs must be tightly controlled to preserve profits. This means that, if a project is going to be completed on time and earn a profit, the project schedule must be developed in a tight and error-free manner. There is no longer any room in the schedule for contingency extras or absorbed costs.

Accordingly, because the job of project scheduler is not standardized but changes from project to project, this reference manual and integrated software program have been compiled from years of collective professionals' experiences to provide you with the type of flexible controls you need to accomplish your project scheduling and bring that project into existence on time, within budget, and by using computerized CPM, at a professional level of technical quality. To provide professional reference, this reference manual is structured in the correct sequence to reflect real-world job precedence of professional project scheduling.

For, at the real-world application level, events never go as simply or easily as they appeared to on the drawing board or computer program. At the field production stage, good project scheduling equates to how closely your planning of all the tasks and activities of the project's development relates to the real-world application tasks involved in making it happen. The owner is always going to look at you and think, is the project on schedule and production going well? Is your scheduling of the related construction activities adequate? Are the production problems that invariably pop up being solved quickly by this project scheduler? Is contingency planning needed but was overlooked? Are the production milestones being reached soon enough to protect my investment?

All this is important to keep in mind, because each project is a unique creation with contractual start and finish dates, and a strict cost budget that must be adhered to realize any profit, and to avoid litigation.

The cost overruns and completion date delays that were so pervasive in the past are no longer tolerated in any production environment. Companies that don't tighten up efficiency and modernize their operations to meet the new construction industry paradigm will not be around next year.

But these were not the full extent of the problems that led to the creation of career project schedulers. In addition to the aforementioned deficiencies in the old ways of construction management, a traditional problem has been the organization of project teams. Project teams typically are made up of people who have limited or no experience in working together as a team. This may be their first time working together, and possibly the last. Or perhaps there are hidden problems from the past among these people. One never knows. So these constraints that we do know of make project scheduling very difficult, and those constraints that we are unaware of behind the scenes, can sometimes make your job similar to tap dancing in a rattlesnake farm. Yet it can be very profitable for the professional CPM scheduler skillful enough to do it. These are professional project schedulers who have learned to manipulate time. And in the construction industry, time really is money. Control the time and you control the profit.

All construction projects, be they residential, commercial, industrial or public works, are complex to some degree, with larger projects being monumental in terms of the amount of the activities and tasks involved. Industrial projects involve thousands of activities and hundreds of contractors and suppliers, on multiple critical paths network scheduling. For a project to be completed on time and on budget, all work (which is broken down into tasks, activities, then phases of completion) must be carefully planned and scheduled in advance. The professional project scheduler then "cooks" the schedule with integrated computer software programs, looking for any long-lead items and all production bottleneck potentials. If all the activities would simply follow each other in consecutive order, the job of project scheduling would be easy enough for the owner to handle without hiring a professional project scheduler. In smaller residential jobs, scheduling is part of the general contractor's job. However, even "small" residential construction contractors get busy when many bids come through at once, and they can soon be swamped with workload. For the bigger jobs to be contracted profitably, a good CPM project scheduler is a necessity.

Because of the complexity of modern construction techniques, and the high degree of specialization and licensing within construction management, owners frequently retain professional project schedulers to develop and operate the project's production schedule. The experienced project developer knows that licensing and registration laws require certain contractual portions of the project to be done by persons possessing designated licenses. And that other portions of the work will be delegated to other employees of that firm, or outside consultants retained by the project management firm.

Project schedulers are generally considered personal consultants hired by either the owner or project team. A client who retains a scheduling professional does so because the client is impressed with the professional skill of the scheduler, or the firm that scheduler represents. Today's project developers generally specify that specialized CPM scheduling work be done by highly qualified specialists, often hired as consultants beyond the parameters of the project team.

Profit margins in the construction industry during these last lean decades have shrunk from twenty percent to three percent, forcing contractors to take on more jobs to make the same annual income. Because of this, CPM project schedulers have seen an increased demand for their services from contractors as well as owners, a modern window of opportunity due to high competition for a low margin. This also will continue well into the foreseeable future. The old days of many contractors relying on Time & Materials extra work orders from fat, drawn-out projects are long gone. And today's commercial projects are now unfundable without CPM production scheduling.

Since each project is unique, the production tasks and activities will be vast and different with each project. It follows then, that CPM network scheduling is uniquely complex for each project. However, the SOP for network scheduling detailed within this book are the fundamentals upon which all successful project scheduling is based. Your adaptation of these procedures, and mastery of the accompanying software, will provide you the foundation of modern CPM project scheduling and the career skills necessary to pursue a profitable career in CPM project scheduling.

Scheduling Fundamentals

A project schedule is the contractual network diagram of the project's planned activities, their sequence determined by job logic, the contractual time in working days required for completion (activity duration), and the conditions necessary for their completion (contract specifications). It is also a contract document linking the lender, developer, prime contractor, and subcontractors. It serves, within the contract specifications, to advise the lender and developer of any unsatisfactory progress in any activity's production, and as a strategy to the prime contractor and activities subcontractors of the jobs they must accomplish within their contractual time frame. It is accepted industrywide that use of CPM network scheduling in modern construction projects is the difference between budget success or bankruptcy. CPM has proven itself to have these three distinct fundamental advantages in project scheduling:

- CPM scheduling provides instantaneous monitoring of the project's production.
- CPM scheduling increases productivity and efficiency of the project's production.
- CPM schedules are contract documents that stand up in court to prove or deny contractor delay claims.

Now let's first back up a bit and start at the beginning, by defining project scheduling. In the conception stage, project scheduling is the process of carefully considering all the activities of the project's production. We group them into phases, list them, then graph them. This is then followed by the planning stage, whereby one determines the steps required to accomplish each activity, then lays out those steps in a logical production sequence. We call this *job logic*. Many things influence job logic. For example, activity trades subcontractors availability, fabricated items' delivery dates, or perhaps a trucking strike could delay long-lead items' delivery at the last moment. So if delivery in your area could be a potential problem, the delay would delay other activities and thus influence the job logic or activity sequence for your area.

Next comes the scheduling phase, using the fundamentals of CPM project scheduling to coordinate the many tasks, activities, and phases necessary to bring that project into existence. Then the numbers are crunched and the schedule is cooked, which means to apply critical analysis to debug the system. When the schedule is authorized by the lender and owner the project begins, and the project scheduler now starts doing cost tracking and audit trail procedures. The professional project scheduler now also begins doing quarterly trends analysis, to see how the project schedule will do next quarter and to determine what changes are needed now. When the client asks the scheduler, "How are we doing?" the professional scheduler has the answers for both now and the next quarter. Both the problems *and* the solutions.

So in this secondary analysis stage, scheduling fundamentals begin by the proposed schedule being broken down into its fundamental components to look like this:

1. You figure out who's going to do what, then you assign a name, number, and the responsible subcontractor to each activity.
2. When are the activities' start and stop dates? How much of the total project will be completed when they finish? Which activities are critical to total project completion?
3. Where in the schedule do their tasks and activities fall? Any slack time or leeway between their starts and finishes?
4. With whom do their activities interact? Link those subcontractors in job logic.

5. What are the activities that will take the longest to complete?
6. Identify all long-lead items (items that have to be fabricated or custom manufactured with a long delivery time).
7. Identify for close monitoring all subcontractors with long-lead items in their activity.

The crucial part of successful project scheduling, however, is the methodology you use to complete these tasks. Profitable project scheduling requires knowledge of network scheduling and the use of modern computer CPM programs. The traditional old ways of handwritten spreadsheets and day planners are long antiquated, although there are still scheduling companies around that offer hand processing for certain types of projects. But, the computer by far, is today's professional scheduler workbench. And integrated software products configured with dedicated program logic are the scheduler's tools. The accompanying software program is an example of integrated CPM dedicated software. You must prepare for tomorrow's profit by mastering the technology of today.

Professional project scheduling always contains a dichotomy, or duality, that is a strange combination of clean, conceptual, theoretical planning which then must be carried out in a real-world field application production that is sometimes much akin to a fistfight in a mudhole. Seldom are the two conducive to easy production; not for anything larger than, say, a doghouse. (Unless, of course, the doghouse is located in a flood plain or earthquake area. Then engineering plans, environmental impact studies and building permits will be required.)

But therein lies the job description for the professional project scheduler: Take the owner's project that looked good enough on computer to get the bank to fund it and develop a production schedule that will build it profitably in the real-time world amidst heavy competitive business combat. And then, of course, there are always the hidden factors in the business world, such as behind-the-scenes egos and power plays revolving around money, that one must learn to manipulate. Although you will never see that last sentence on a job application, you must be adept at these business survival skills also.

Currently, what is evolving in the construction industry is a situation wherein the project manager is working 14-hour days, handling field reports from site supervisors, overseeing contractors, managing activity operations, and interfacing with regulatory agencies and the project team as the owner's agent. More and more, a separate project scheduler is hired to take on responsibility in developing the production schedule, overseeing same and reporting progress to the project manager, owner, project team, prime contractor and subcontractors through specialized summary reports for each, called *sorts*. Obviously this means more work, longer duration of jobs, and more annual income to the astute CPM project scheduler. Today's hot project scheduler also negotiates for a healthy percentage of profits if he or she brings it in under budget, before deadline, or both. This type of incentive profit sharing is a sizeable source of potential income for the professional project scheduler. I would imagine that it is probably a great deal more gratifying to be a project scheduler who has negotiated a piece of profit sharing when the project comes in under budget, than to be one of those poor schedulers who has not yet stepped up to CPM in their network scheduling. I can't quite be certain, as I only work with the former and it is as stimulating as anything you can do with your clothes on.

Modern project scheduling has evolved from traditional monthly planning calendars that were used in combination with handwritten data spreadsheets, and is based upon two types of flow charts, called bar charts and velocity diagrams. In scheduling fundamentals, we refer to velocity diagrams as *S-charts*. They show the established schedule of total tasks and activities graphically as well as record progress along the way. Traditionally, they have proven themselves to be the most useful and functional methods of graphing the project schedule out visually, with hand spreadsheets handling the data flow for summary reports.

But bar charts are inadequate for network scheduling, because of their lack of phase interrelationship float (scheduling leeway or slack time) capability. This results in a lack of control over the critical path of the total work to be performed. Traditionally, this shortcoming was filled in by the prime contractor's experience and intuition. Naturally, one can intuitively grasp that this kind of situation will quickly get out of control with more contractors added to the formula. And lack of control always equates to lack of profit. The S-curve chart is a better scheduling tool because it shows the various activities linked by their interrelationships and tasks' interdependence existing throughout the various sequential phases of the project. Although the S-chart is a better system for scheduling smaller construction jobs and linear time-scale projects, it does not provide time-scale manipulation. And time-scale manipulation is the control domain of CPM project scheduling.

As you will see by using the enclosed CPM integrated software, computer programs are much more efficient scheduling systems because of their instantaneous mathematical calculations, error-free, that "wash through" adjustments in other critical areas, like float and phase interactivities. CPM software systems are configured with program logic, further providing the critical analysis of time-scaled activities sequence by interaction on network diagrams. The analysis is done by critically evaluating the summary reports of these interactions in the program's sorts. When network diagram software systems, such as the one enclosed, are used in conjunction with bar charts, the resulting CPM management control system is the basis of modern project scheduling. Currently in the construction industry, CPM control is a contractual requirement, usually appearing in the contract specifications in virtually all commercial, multi-residential, and industrial construction. And although CPM was designed for the construction industry, it functions extremely well in other businesses that deal with production, such as product manufacturing, distribution centers, and the semiconductor industry.

CPM scheduling fundamentals require that each activity within the project have its own time frame requirement, and usually cannot start until preceding activities have been finished. There may be other activities, however, that can be carried on simultaneously because they are entirely independent of one another, such as plumbing and electrical. "Dovetailing," or interlinking, these activities and their duration events efficiently requires skillful time management planning in the project schedule. A typical construction project involves many varied activities, and tasks within those activities, that are interdependent with one another as well, as other activities that are totally independent of each other. When the production phases, showing all the activities, are interrelated on a network diagram, they create a multi-lane highway of activities moving towards completion. It also shows their individual time durations, as well as their relationships with their preceding and succeeding activities. When all the project's tasks and activities are superimposed to provide the big picture, it quickly becomes evident to the observer that CPM project scheduling is a difficult and complicated project management function. It has traditionally been the project manager's responsibility to oversee and evaluate prime contractors' construction schedules to determine whether the subcontractors are finishing by their contractually required dates. However, this responsibility is now falling within the project scheduler's domain, as construction profits become more closely tied to time-manipulation techniques.

Chain Of Command

A wise instructor of mine back in college once told me, "If you want to learn how any business works, just watch where the flow of money goes." And sure enough, money is the flow-line yardstick within the construction project's chain of command. Any person or organization with a financial interest in the project is called a *vested party*. Any contractor who works on the project has *lien rights*. The first, or primary, level in the chain of command are the vested parties in the project and the top slot here starts with the generator of funds, usually the bank. This entity financing the project heads the pyramid. Next comes the owner and the project team consisting of the designer or architect who drew up the plans, followed by any registered engineers who were consulted on a commission basis regarding structural design analysis. These primary parties, such as the architect or engineers, may have subcontractors or consultants who would be junior to them in authority. Next comes the project manager, who may be one of these engineers or may be a construction management professional from another field. These parties may also have subcontractors or consultants working for them. Then comes the project scheduler, followed by the site superintendent, and then field personnel. The project team's staff finishes out the primary chain of command.

The secondary level of chain of command in the project includes the workers who improved and built the project. These are lien holders, starting with the general, or prime, contractor, then followed by the subcontractors. Subcontractors are then followed by their equipment and material suppliers. These parties are also vested in the project by their lien rights, which provide for legal grounds of foreclosure in the event of nonpayment. In essence, the subcontractors and suppliers are extending credit to the project and hold what is legally termed "a cloud on the title," which are foreclosure rights in the event of payment default, as with a mortgage company. We will cover owner - contractor relationships and construction project lien rights in greater detail in Chapter 9.

Business being what it is, disagreements often emerge between these vested parties and lien holder parties. Construction litigation courts are full. The business combat chapter, Chapter 8, delves into contract dispute resolution and the methodology for keeping your schedule intact as unforeseen power plays emerge among the chain of command as the project progresses. Procedures for keeping contract documents tracked on your schedule are covered in the software chapter, Chapter 15.

Chapter

2

Scheduling Systems

Systems Fundamentals

A project schedule becomes operable when the scheduling system has a functioning network diagram of the activities. A working flowchart with all job logic sequence developed, and total durations required for completion added, creates a network timeline. A schedule's primary purpose is to keep the project on schedule. However, much useful data from the schedule can be used in other ways for added cost-efficiency in many areas of project production. As we have seen, computerized CPM increases productivity, efficiency, and time management, thereby reducing overall project cost. The fundamentals of the system can be illustrated by a simple project, such as a room addition. The activities could be as simple as starting and finishing dates for the foundation, walls and roof. A more complex project, such as the entire residence, will add phases of activities completion. These phases serve as benchmarks in construction completed to date and according progress payments to the activity subcontractors. In systems fundamentals, these benchmarks of phases completion are called *milestones*. Financial milestones are typically tied to the dates progress payments are due.

Systems fundamentals of project scheduling came into being on capital projects in the early forties, with preplanned, written schedules. During World War II, the military began to develop their own type of project scheduling to deal with multi-project management of many different types of project contractors. The first is an activity duration estimate system developed by the military, called PERT, which we will examine later. In the civilian market, two traditional systems have evolved from hand spreadsheet methods: bar charts and logic-diagram-based schedules. Both methods are used extensively and interchangeably in both project planning and actual construction work. Naturally, each method has its advantages and disadvantages.

Bar charts are the simplest form of scheduling and have been in use the longest of any of the systems we have available today. They offer the advantage of being cheap and simple to prepare, they are easy to read and update, and they are readily understood by anyone with a basic knowledge of the construction project business. They are still in wide use today, even as one of the final sort reports of the computerized CPM scheduling system. The main disadvantage of the bar chart is its inability to show enough interactive relationships between all of the activities on larger, more complex projects.

Figure 2-1 is an example of how visual data overload doesn't allow for a clear path, made up of many smaller paths, to become apparent. In the bar chart illustrated here, there is no critical path method of scheduling. Simple job logic of placing the activities in order of their construction sequence was used. Note how the many activities have been condensed into single group bars to get the complete chart to fit on one page. Any time you condense something by compressing it into something else, you lose control over that item. In network scheduling, the approach is just the opposite. We "explode" the activities, subactivities, tasks, and work items out into a network so that each has complete computer tracking and controlling. Each of the macroactivity bars in Fig. 2-1 need to be broken down into a series of individual tasks, to control the microactivities within the macroactivity. Only by having control at those levels will we have ultimate control of the macroactivity.

The bar chart here shows macroactivity progress, which is useful and will suffice on small jobs, but no interrelationship between the activities that would allow for alternate routes if progress hits a bottleneck. This ability to switch paths is crucial to Critical Path Management. It's much like tap dancing in a minefield. You can't stop moving or you lose. If project production stops, everyone loses, so it's in everyone's best interest that things keep moving to a timely completion. Bar charts, such as that shown in Fig. 2-1 have no capacity for contingency planning. And your job as project scheduler is to be ready to pull the rabbit out of the hat by having a workaround ready to go if the schedule encounters a bottleneck. So this type of bar chart is useful but it is not enough.

This example has excellent comparison and completion capabilities, and bar charts have a number of other very useful methods of processing data. Note that some of the activities finish ahead or behind where they would in smooth progression. This is vital information and must be tracked accordingly, but no option exists in Fig. 2-1 for critically analyzing the activities interrelationship, or doing something about them. CPM offers those options and, if float is used correctly, also offers the time management necessary for critical path workarounds. Our example here shows the strategic dates for starting and finishing major portions of the overall project, along with a few milestones. Intuitively, one can grasp that there is little or no ability to control changes or manipulate time-scale interlinkage between activities in such macroactivity grouping.

Control comes at the micrograuping, or individual task, level. Manually generated bar charts and spreadsheets that cover the whole project involve a lot of detailed record keeping and summary reports, which soon absorb too much of the project manager's time to allow managing the field end of the project's production work properly. This means, in our example, that each bar would also need another series of bars to show subactivity tasks within each activity noted. For example, under the contract phase, the Supplementary

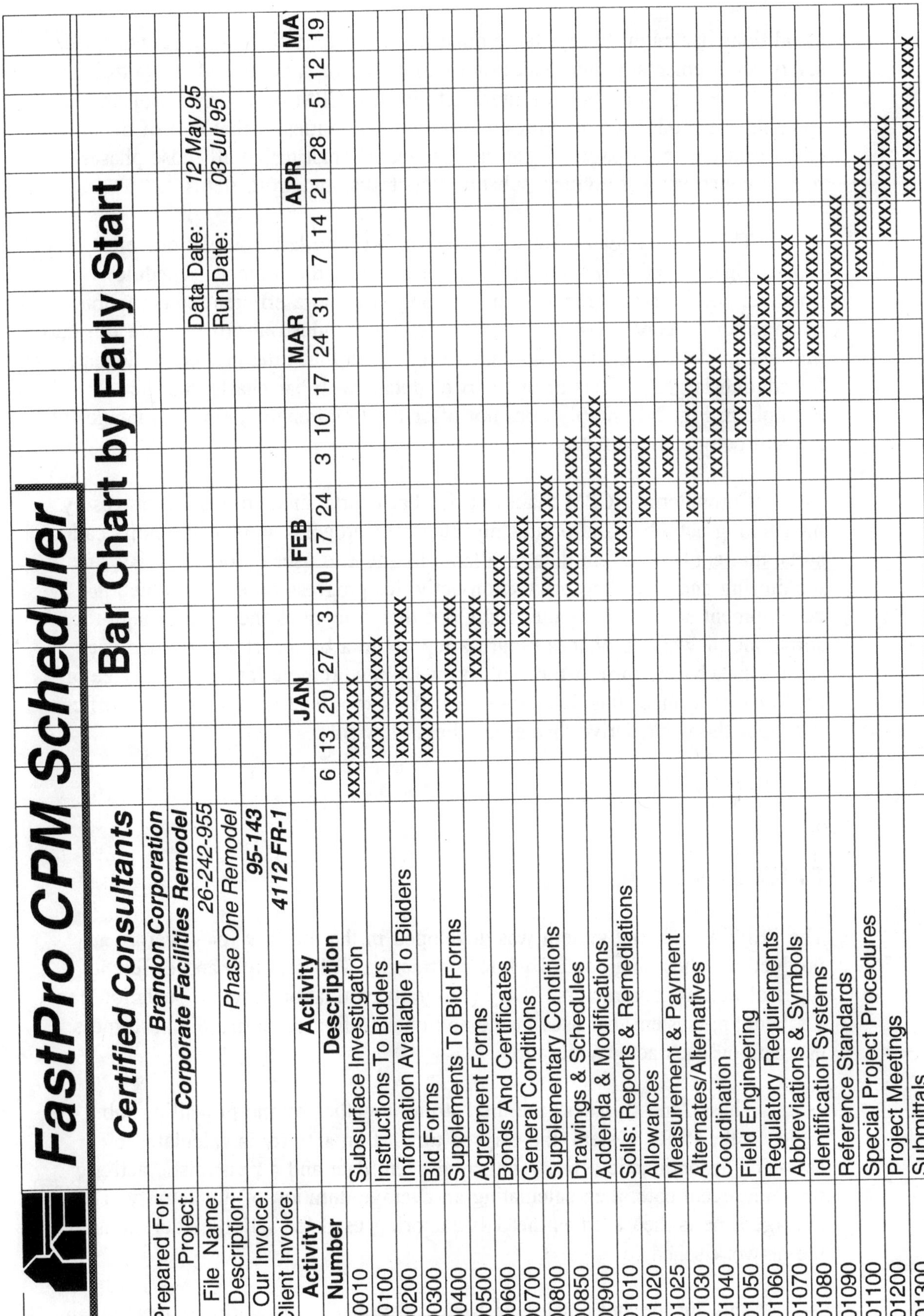

FastPro CPM Scheduler

Certified Consultants

Bar Chart by Early Start

Prepared For:	**Brandon Corporation**
Project:	**Corporate Facilities Remodel**
File Name:	*26-242-955*
Description:	*Phase One Remodel*
Our Invoice:	**95-143**
Client Invoice:	**4112 FR-1**

Data Date: *12 May 95*
Run Date: *03 Jul 95*

Activity Number	Activity Description	JAN				FEB				MAR					APR				MA		
		6	13	20	27	3	**10**	17	24	3	10	17	24	31	7	14	21	28	5	12	19
00010	Subsurface Investigation	XXX	XXX	XXXX																	
00100	Instructions To Bidders		XXX	XXX	XXXX																
00200	Information Available To Bidders		XXX	XXX	XXX	XXXX															
00300	Bid Forms		XXX	XXXX																	
00400	Supplements To Bid Forms			XXX	XXX	XXXX															
00500	Agreement Forms				XXX	XXXX															
00600	Bonds And Certificates					XXX	XXXX														
00700	General Conditions					XXX	XXX	XXXX													
00800	Supplementary Conditions						XXX	XXX	XXXX												
00850	Drawings & Schedules						XXX	XXX	XXX	XXXX											
00900	Addenda & Modifications							XXX	XXX	XXX	XXXX										
01010	Soils: Reports & Remediations							XXX	XXX	XXXX											
01020	Allowances								XXX	XXXX											
01025	Measurement & Payment									XXXX											
01030	Alternates/Alternatives								XXX	XXX	XXX	XXXX									
01040	Coordination									XXX	XXX	XXXX									
01050	Field Engineering										XXX	XXX	XXXX								
01060	Regulatory Requirements											XXX	XXX	XXXX							
01070	Abbreviations & Symbols												XXX	XXX	XXXX						
01080	Identification Systems												XXX	XXX	XXXX						
01090	Reference Standards													XXX	XXX	XXXX					
01100	Special Project Procedures														XXX	XXX	XXXX				
01200	Project Meetings															XXX	XXX	XXXX			
01300	Submittals																XXX	XXX	XXX	XXXX	

Figure 2-1
Bar Chart By Early Sort

Conditions bar might have a large quantity of subactivities covering workplace safety requirements, public pedestrian barricades, or specific traffic control plans. These would all be individual activities in CPM on a larger project. But showing the production activities and their subactivities in that sort of detail could easily take 20 pages of bar chart schedules just to list all those phases-related construction activities, subactivities, tasks, and work items.

The primary reason for the cost overruns and late completions of projects in the sixties was this lack of control over the project's schedule timeline, and became a focal point of change for more efficient construction management. Delay overruns on commercial and industrial projects can run into millions of dollars per day. Bar charts used up to that time proved inadequate. Scheduling larger projects in that sort of detail, with bar charts like the example in Fig. 2-1, simply does not offer the time-manipulation advantages of modern computerized CPM.

Computerized CPM makes up for bar charts' inherent disadvantages by interlinking activities and producing integrated progress summary reports called sorts, that contain current data, available to any end user on demand. By first monitoring and recording production activities progress on each production development activity, computerized CPM then integrates the data through the sorts, one of which produces progress reports in a bar chart sort. Experts are saying that, by the turn of the century, we'll have remote terminals the size of credit cards to input this data from a car back to a host computer in the office that does the work. Drive-thru scheduling... I wish!

PERT Scheduling

The PERT scheduling system was developed in the military for government production scheduling control while contracting with multiple and different types of defense contractors. PERT (Program Evaluation Review Technique) is a scheduling system that uses inside and outside figures to make a best-guess estimate on each activity duration.

These "guesstimates" are then strung together by the prevailing job logic. In the PERT system, the elapsed time of an activity is calculated by assessing an optimistic activity finishing event date and a pessimistic activity finishing event date, then calculating an average duration of the activity. This average time is then used as the only factoring used for activities durations in the network schedule.

This "inside and outside best-guess estimate" will work for averages in data processing, but PERT serves no other scheduling functions in modern construction project management.

Accordingly, most modern CPM software programs for construction projects do not include the PERT-type scheduling system, but instead use the data from precise activity events, both early and late, calculate with all four of those factors, then average the duration. Float can also be taken into account. The proper selection of the elapsed duration times for the work activities is an important part in preparing an accurate CPM network schedule, because exact events determine the exact activity duration times, which in turn directly affect the critical path and succeeding activities float times. The PERT system does not offer these scheduling advantages.

In the PERT system, generalized time estimates for activities throughout the total project are made, with timing leeway built in cushion allowed for optimism or pessimism of activities duration. However, the linkage of activities and establishing a critical path through a PERT system is very difficult to establish and impossible to maintain workaround options in the event of changing critical paths. To be accurate in event time estimating, the project scheduler breaks out all activities into manageable units of time. The elapsed-time estimate is based on the unit labor and materials quantities taken off the project's specifications, which are used to estimate the project cost. This quantitive survey estimate contains a wide variety of bidding pre-estimated information, such as cubic yards of import fill or excavation, tons of structural steel, numbers of block masonry units, and cubic yards of concrete.

That information, combined with previously recorded labor and material costs from the company's own records or activity subcontractor, is used to calculate the expected elapsed time duration for each activity. The unit number to be used a computation factor is then adjusted for delay by factoring in prevailing local conditions, such as seasonal inclement weather, variations in delivered materials costs, area union labor rates, trucking strikes, etc.

Note the way that etc. just hangs at the end of that last paragraph. Looks nice and easy, and it sure was easy to write. Just dropped it on the page. But in real-life application, any etc. in that last paragraph will cost thousands of dollars of unrecoverable lost profit. Part of your job is to know and account for the unpredictable events that will delay your schedule, such as inclement weather or materials shortages. Just because it was ordered doesn't mean you're going to get it. Follow up on every line item is part of your job as a professional project scheduler.

Scheduling Philosophy

The dichotomy between scheduling planning and real-world construction execution are the two main factors affecting the preparation of the project schedule. Practical limitations of the project's production must be established, then prepared for in the planning stages. We call this conception and critical thinking stage *scheduling philosophy*. Once the project commencement occurs, the basic ground rules for its construction schedule must have been already established. These will address the scope of construction phases, materials procurement procedures, contingency critical paths planning, and long-lead items procurement. These ground rules will be different in specifics with the job logic of each project, but large or small, they will all determined by following basic scheduling philosophy. Scheduling philosophy refers to the selection of the scheduling system. The questions that must be answered prior to the first scheduling meeting will determine the scheduling philosophy, which in turn will determine the production contracting basis. Scheduling philosophy uses critical analysis to determine:

- The scheduling management needed for CPM
- Identification of critical path activities
- Workarounds (alternative option activities that can be substituted if a critical path activity hits a bottleneck)
- Job logic (sequence of activities)
- Timing of schedule recycling
- Contractual requirements for progress reporting from field supervisors
- Assignment of CPM-trained personnel to areas of responsibilities

Typically, it is the project team's responsibility to establish the scheduling philosophy, critically analyze their decisions, then get owner approval of scheduling system selection. The scheduling philosophy must be a cohesive decision among all vested parties because changing the system later during project's phases is disruptive of the production continuity, and accordingly, very costly. The scheduling philosophy should also account for the start-up sequence of the various activities making up the production phases. Activities that start up first must also be the activities that finish first. The project schedule must also interlink with the other major project considerations, such as design and long-lead items procurement. That is easy enough where, traditionally, the design is completed before the project goes to construction bidding. But in other production execution modes, such as fast-tracking, design should be 30 to 50 percent complete if a reliable CPM schedule is to be developed and executed. By getting this advance start on a project without a finished projects design, error is obviously probable if the operation has not been done previously to provide accurate activities duration.

Field Scheduling

Even a detailed CPM schedule with a comprehensive work breakdown structure is not an adequate tool for controlling the daily progress of the activities in the field. Field scheduling is necessary to coordinate reporting data with the main scheduling system. On smaller construction projects, the production progress milestone schedule may be the only one used. Sequential completion of phases is simple enough on small jobs and usually will be adequate. On larger projects, however, the project schedule also functions as the basis for making more detailed weekly work plans in the field, for each major activity. In either case, the approved project schedule will, in turn, determine the type and detail of the field scheduling used by the owner and the prime contractor to set the production phases milestones and monitor the projected schedule progress versus the actual completed progress. Detailed field scheduling is necessary continually throughout the entire project production to interlink the two for critical analysis and control.

Complete detailed field planning makes the most efficient use of field manpower on the priority list of tasks within those activities that are required to meet the CPM milestone dates. This detailed planning first concentrates on those items of work that are on the critical path, then on secondary activities that are non-critical. The field scheduling personnel of the project team are responsible to the project manager for planning the construction activities each week for the following week. Field schedulers list out the work activities sequentially that are scheduled to be completed by the following week. This list of pending activities is then discussed in the weekly scheduling meeting, with the project manager, project scheduler, prime contractor, major subcontractors, the chief field engineer, and the site superintendent all present. These weekly meetings are typically chaired by the project manager and/or the project scheduler.

Prior to the meeting, the site supervisor will have checked to see that all the necessary labor, materials, and equipment are on site to perform next week's scheduled work. The site supervisor will then report production readiness or deficiencies at the meeting. After the meeting, the weekly task lists are distributed to all responsible personnel, so the activities subcontractors can plan weekly production execution. Scheduled work not completed in the previous week is also examined at the meeting, and plans are made to bring those activities up to speed. The outlook for the next two weeks is critically analyzed and projected. The project scheduler then prepares the longer-range quarterly field scheduling plans by checking the critical CPM milestone dates four weeks ahead, and plotting the appropriate course(s) for the next quarter.

Proposal Scheduling

The project schedule first comes under serious discussion by the contractors when the bidding documents or the request for proposal (RFP) arrives at the individual contractor's office. The RFP or bidding documents contain a section devoted to the construction schedule. Usually the documents ask the contractors to develop preliminary bar chart schedules showing the major construction activities (and their durations) in the contractors' proposals. The purpose of that request is to get the contractors to look at their strategic end dates in general terms to see if they are still feasible. The owners use the contractors' preliminary bidding schedule to compare it with the schedule the project team has developed. This is typically the project team's first opportunity to get the activities subcontractors' professional opinion as to how long the activities' construction work will actually take. This is then compared to projected scheduled duration and adjustments made accordingly.

Now, the contractors are not paid for doing this; it's just part of the business. So typically, contractors don't like to spend much time developing construction schedules during the bidding process. At this point they are not assured of getting any contracts. The only exception to this reluctance of contractor input is when the selection of that activity contractor may depend on a previous track record of being able to meet a tight completion date. This is normally the case in fast-tracking project production execution. Then the contractor's schedule becomes a bidding edge that is worth the extra bidding expense to the contractor. In the case of fast-tracking projects, the contractor's proposed schedule will undergo intensive examination by the project team during bid openings, and must be viable within the owner's project schedule.

At this proposal stage, the project scheduler must evaluate the schedule in relation to the contracting plan and the construction technology available to improve the production timeline. For example, there could be company policies affecting the contracting plan that could be changed for this project to improve this particular project schedule. If they are, the new suggestions are then reviewed by both the owner's project team and the prime contractor to determine if the proposed changes are viable. This is the same procedure for substituting construction technology improvements that will shorten an activity's completion duration.

As a professional project scheduler, you must be aware of where the "hot spots" are that will make you or break you. Proposal scheduling is a critical juncture of one of the hotter spots. Here you must show your experience and critical analysis to the project team. At this point, project schedulers must make their opinions known to the owner, client or project management team if they disagree with what is being developed in the proposal scheduling. The next stage is the schedule going into contract, so not voicing any intuitive concerns at this juncture will appear to others as consent to the proposed terms and conditions that will be going into the contract as contractual obligations. Speak now or forever hold your peace.

Scheduling System Selection

The selection of the scheduling system we are going to use in running our project's production is understandably crucial to overall success. To be cost-effective, the selection must consider the following criteria:

- Size and complexity of project
- Scope of services required
- Sophistication of users (owner, project team, field personnel, subcontractors, etc.)
- Available scheduling systems
- Owner preference
- CPM scheduling costs versus savings
- Contract schedule specifications

Your company's experience with its present systems is an important factor, because using a system with which your people are familiar, that has proved itself reliable on similar projects improves manpower time-efficiency in operating the system. Introducing a new system on a project already in progress causes more problems than it solves and is often the kiss of death. Accordingly, the curve of cost-effective schedule control usually goes *down* upon initial introduction and use of a new CPM system, until the system becomes comfortable and familiar to those who use it. It follows then, that the time for training personnel in the use of any new system is *before* operations commence.

The minimum hardware requirements for CPM network scheduling are an IBM-compatible 486 PC of 640K or better memory, an 80-megabyte hard disk or better, and a laser printer. Presentation graphics require a plotter, but that is not essential to doing a decent job of schedule control. The cost of the program's training is the hidden factor in the scheduling cost budget, and this depends on the experience and computer literacy of the project team. But regardless of the cost, training is crucial to the success of systems implementation. If a CPM system is not run by CPM-trained personnel, it will not be cost-efficient.

The cost-effective savings in using a computerized schedule comes from the generation of much more relevant and usable data than is possible with bar charts. Data generated by computer has a lower unit cost than labor-intensive, hand generated data. However, if that computerized data is being processed by untrained personnel, the system is not being used to its fullest advantage and will not prove to be cost-effective in the long run.

Integrated Systems

Computerized project management consists of integrated project controls. These include planning, scheduling, and performance-measurement aspects. These controls have to encompass the basic business parameters of progress payments and time billing, computer assisted design (CAD), cost and time estimating, cost databases, job cost, equipment and inventory costing, procurement, contract document control, and administration. All these facets must be integrated to "wash through" automatically to the primary levels of the project's accounting, which are AP (accounts payable), AR (accounts receivable), and PYRL (payroll). Even though your project schedule is only one factor in the entire project process, you can see how important it is and how it will influence all the other factors. So just being aware is not enough. You must understand the link between the project schedule and costing areas to see the big picture and appreciate the pivotal nature of your project schedule. A fully integrated CPM system will have most, if not all, of the following components:

- Computer Hardware and Peripherals
  Computers, scanners, digitizers, monitors, printers, plotters, video, CD ROM, telecom linkage modems
- Connectivity
  LAN (Local Area Network), WAN (Wide Area network).
- Accounting & Job Costing Program
- Estimating Program
- Contract Document Control Program
- Procurement Program
- Planning, Scheduling & Performance Measurement Milestones
- Computer Assisted Design (CAD)
- Network Timeline Scheduling Program

Today, with personal computers tied together in local area networks (LANS), schedulers can completely replace expensive and cumbersome mainframe and minicomputer project management systems. Modern CPM software provides the ease-of-use of personal computers while improving communications among project team members who need instantaneous access to project data, files, and summary reports. Project scheduling data stored on file servers can be made accessible to any member of the project team who needs data retrieval immediately. This instantaneous data retrieval for several end users is termed *multiuser server functionality*. These modern types of systems provide controlled concurrent access to project files, so that the entire project team has select capabilities for scheduling, resource management, cost control, and summary reports (sorts).

Project team members aren't always in the same office, on the same floor or even in the same town. They may even be in different states while the project work is taking place. They don't necessarily work the same hours or schedules, may not be in the same time zones, and aren't always available by phone or modem when some other member of the project team needs them. But by using a multiuser LAN system they can get in touch at a regularly scheduled time, and can stay up-to-date regardless of their individual work schedules or geographic locations. The key to any schedule's success (beyond activities duration control and phases time management) is consistency, through common reporting specifications and sort formats that are understood and workable by all responsible project team members.

Systematic reporting to those who are responsible for production, is essential. When time-scale interlinking of production activities must be changed to reflect a new critical path, multiuser server functionality is the type of sort system that fills the bill. Combined with a software tracking system, the scheduler can then provide computerized audit trails of budget expenditures to date, and cost tracking reports to analyze how effectively that money was spent.

Chapter

3

Critical Path Management

Nomenclature

Professional project scheduling is built on the fundamentals of the Critical Path Method, or CPM as it is known in the trade. Critical Path Method refers to identifying bottlenecks in the production process and then building a production timeline through them, using the total duration of those activities which will take the *longest* to complete as the basic timescale. Critical Path Management is the methodology for managing those timeline paths efficiently, by manipulating time management of the interlinkage of activities durations and contingency workarounds. It is a scheduling system that allows the project scheduler to achieve improved time control over project's production phases. Additionally, an audit trail and cost tracking can be incorporated into the system to provide a reasonably accurate estimate of timeframes required for those or similar activities in future projects. CPM typically works best in straight-line, time-scaled productions in any business where the timelines can be estimated with a fair amount of certainty. These network scheduling systems are essentially a combination of CPM and S-charts. The accompanying software program is an example of an integrated CPM program system.

Critical Path Management is the most accurate computerized system of network scheduling yet developed. It allows production managers, developers, owners and prime contractors to achieve control in the following critical areas of project scheduling:

- They can figure out where they stand right now in the project.
- They can determine where to expect production bottlenecks.
- They can decide what to do next.
- They can begin changes today to keep the project on track.

In the construction industry, repeat business does not just come from building something well. The companies that get repeat business do so because they manage projects in ways that protect their clients. CPM scheduling tells the owner where he stands, not only financially but also in terms of issues, problems, and resources on the project. Ultimately, that's the biggest concern to most owners. CPM has proven itself by adding more bottom line to owners' investments. In larger projects, such as multi-residential, commercial, industrial and public works, CPM is mandated by contract.

All production activities will affect each other in either direct or indirect linkage. CPM fundamentals require that a critical activity must be finished before the succeeding critical activity can start. Non-critical activities and those of similar trades, such as roughing in the plumbing and electrical work, can proceed simultaneously. One of the objectives in Critical Path Management is to find the order in which the phases of critical activities must be completed. Those designated as critical activities begin by being assigned to the production activities that will take the longest to complete and are crucial to the production timeline. By tracking these activities, attention is automatically drawn also to those areas where it is most essential to avoid production delay. When bottlenecks occur in the schedule, those activities and their subactivities on the critical path must be handled first, while those delays off the critical path can be addressed secondarily. This is especially true if a particular activity has zero or negative float. Delays along the critical path begin affecting total project duration. Critical paths may change as production bottlenecks are broken, critical subactivities are changed, or when new problems surface.

CPM is currently is the only production scheduling system that covers all the phases of a project and allows the scheduler to manipulate timeframes during and around the activities that will take the longest time to complete, then selecting the best ways to expedite or work around the phases of activities for contingency planning. The underlying logic here is that contingency planning reduces risk. When this time-scale interlinking of production activities is combined with a software tracking system, the scheduler can then provide computerized audit trails and cost tracking reports, as well as trends analysis, which is next quarter's forecast of the current numbers' logical outcome. These are crucial functions for protecting your client's vested interests in the project, and are major selling points of your services as a professional scheduler.

CPM can be though of as the path of least float, based on the relationship between sequential activities and completion duration of the activities. We will examine float in greater detail in Chapter 5, but for now, think of float as scheduling time leeway, or slack. In developing a CPM schedule, the critical path is first determined by identifying all activities with zero float time. Any activity is considered critical if its completion delay will cause total project completion delay. Critical path activities with zero float time must begin when scheduled, or the total project completion date will be pushed back. The Arrow Diagramming Method (ADM) and the Precedence Diagramming Method (PDM) are the two basic S-chart scheduling standards upon which CPM has evolved.

One of the cost-effective benefits achieved by using CPM scheduling is the analytical consideration of activities duration and job logic sequencing of each activity in the initial stages of the project as the CPM schedule is being developed. An example of such a CPM network diagram can be described by an arrow diagram. The tail end of the arrow indicates the start of an activity and the head end represents the completion of that activity. Using a graphic arrow to indicate paths on a flow chart, each activity (or arrow) will have a start, stop, and duration. When two or more arrows or activities meet, the intersection is called an *event*. Activities are begun and completed at events, and succeeding activities in their production phases move forward from one event to another event. The events are assigned numbers, which are used by the computer to change event sequences or durations as the CPM program is monitored and recycled during the production period. The various activities and events in the CPM schedule, making up the network diagram, are interlinked by interdependence and the project's timeline.

A CPM schedule shows the interdependence of one activity on a preceding or succeeding activity, much like a shadow. For instance, if the excavation for the building's footings is scheduled to commence ten days before forming activity commences and sixteen days before placing of the concrete event commences, any delays due to late delivery of materials or inclement weather will cause subsequent activities to be delayed, pushing back the contractual completion date. Typically, in modern commercial projects, a structural-steel framing system will be erected on the building's foundation. So if, in our example, the earthwork excavation delayed the forming and pouring of the footings, the delay would also push back the entire phase of activities involving placement of the steel and iron workers. The CPM schedule will also show what effect inclement weather delays will have on foundation work, completion leading to the structural-steel starting events and finishing events, and the relations of these changes to all other construction phases. If there is zero float time on the critical path, the project scheduler using CPM can decide whether the concrete foundation work is to be accelerated or the steel erection can be delayed.

Showing interdependence of one task, or work item, on another is the major difference between bar charts and CPM scheduling methods. Whereas the standard bar chart may show a continuous line of activity for a particular trade, which starts at one point and continues uninterrupted to another point, the CPM chart draws attention to the specific starting and ending dates (events) for each major portion of a construction component. The CPM will also show the interdependence of each activity and the effect of one event finishing late or early on the starting event of the succeeding construction activity.

CPM scheduling has disadvantages, also. It will increase the total contract price. Such schedules are expensive to create and maintain. In addition, a professional CPM scheduler must be hired to develop and manage the schedule. The project team, or at least those who are responsible for production, must be trained in CPM. Finally, small- to medium-sized contracting companies typically believe such a sophisticated schedule is unnecessary and a hindrance to work. In such a case, the subcontractor's developing and maintaining a CPM schedule may be haphazard at best. However, requiring the contractors to construct and maintain a CPM schedule has at least three advantages. First, it requires the contractors to work more efficiently. Second, it gives the owner an instantaneous summary of the actual progress to date of the project. Third, from a litigation standpoint, requiring the contractors to maintain a CPM schedule helps prove or disprove financial claims and change orders.

CPM scheduling starts with the preparation of a network diagram displaying all project planning and construction activities required for the project's completion. The primary work items needed to complete each activity must be identified, located, and lined up. The scheduler then determines which tasks must be completed before each following work item can started. Once the overall length of time is determined by adding all the phases that will take the longest, the job logic order in which these activities must be completed is established. Finally a network diagram, which is an activities flowchart, is worked out.

Some project activities precede others on a straight-line basis, and cannot start until a prior activity has been completed. Other activities can start prior to the completion of a preceding activity, while other activities are performed simultaneously or concurrently with others. Those activities whose durations are crucial to the overall project completion date are considered to be on the critical path. These are the basic factors necessary to construct a CPM network.

Next comes the scheduling phase, in which an estimate of the time required to accomplish each of the activities is developed. This is really an educated "guesstimate," so we have ways of dealing with inside and outside figures for factoring a time cushion, which we call *float*. The third stage involves installing these estimates into the activities network diagram and finally, in the fourth stage, computations are made of data and critical paths of activities to provide the time-scaled network of the project. CPM is traditionally explained as this sequential order for illustration. However, in real-world usage this is rarely the standard procedure. And that's the crux of the matter right there. *There is no standard procedure*. Because each project is unique, no single formula can factor in all the intangibles. The computers can do instantaneous, error-free computations, but it will always take an analytical mind using experienced critical thinking, to program and run those numbers profitably.

For example, in the time-compressed business world of today, these four basic planning steps often are completed simultaneously. Those trained in CPM use programs that "wash through" this data instantly to each stage, and change datum in all related stages if one factor is changed, using "what if?" scenarios. For the sake of instruction, though, we will use the assumption that they are each treated separately and in the order listed above.

Timeline computations are just simple addition and subtraction. The task of time-scaled computation is usually just as simple and easy. However, managing the timelines of the many tasks and activities involved in a modern project quickly becomes an enormous challenge. You will see, as your career progresses, that there are days when your CPM schedule timeline feels like a nervous mustang ready to leap out of control. Accordingly, modern schedulers use computers to manage their CPM timelines. And with the use of dedicated software programs, the timeline can be updated regularly without redoing the entire network diagram. Complex time variables can be "best line linked." Forecasting by use of trends analysis shows probability of various outcomes. Each activity can be updated regularly for accurate data factoring.

Typically, this is done in daily updates. This frequency serves two purposes. First, it guarantees timely accuracy in your data and pries open the largest window of opportunity necessary to make changes and workarounds before problems start costing money. Second, it produces a superior kind of detail orientation, combined with fast results, that tend to prevent budget cost overruns and give the astute project scheduler a definite advantage over the competition. This is the type of professionalism that produces profit and builds the reputation that you are worth a larger paycheck. Your competition consists of scheduling management companies that specialize in all levels of network scheduling, for contractors, production factories, architects, and owners.

The complex scheduling problems that occur in commercial CPM can be worked out rapidly by using program logic in a dedicated software network system, and time management becomes a viable tool for working out options, workarounds, and solutions. A programming task that normally takes hours to complete, if done by hand, can be accomplished in seconds by computer programs incorporating program logic. This rapid access to data is one of the key reasons for the successes of CPM. This speed of program logic allows variables to be worked out or planned around before they cost money. The enclosed accompanying software is configured using velocity network diagramming with program logic.

CPM Terminology

To understand Critical Path Method management, you first must know the lingo of CPM, along with the terms' meanings in network diagram scheduling. Here are the principal terms and definitions of procedures of network scheduling, in order of their precedence:

Task. A task is an individual unit of work that may combine with other tasks to complete an activity or be independent work items. Tasks can be thought of as the separate work unit items that collectively need to be done to finish each activity sufficiently enough to start the next activity.

Activity. Activities are any single identifiable work step in the total project's production. Groups of tasks, combined to finish a job item, are activities. Once the activities have been identified, their sequential task logic is established. Then begins outlining the project graphically into a network diagram. In a velocity diagram, the symbol for an activity is an arrow. The arrow connotes linear timeline movement from left to right, start to finish. The continuity indicator is the arrow from one activity to another.

Activity number. Number assigned to each activity. These numbers should be sequential. The computer will not be able to establish job logic if activity numbers are non-sequential.

Activity list. List of work items for a project; also the work break-down structure.

Activity duration. Elapsed time to perform an activity, start to finish.

Arrow diagram. CPM network diagramming method using arrows to show activities interrelationship and the flow of job logic.

Event. An event is the exact day at which an activity is just starting or finishing. Network program logic applying to all events is that all activities leading into an event can be started at that time. An activity is always preceded by an event and followed by a sequential event. Thus, an activity always has both a starting event and finishing event. Theoretically, that finishing event is the starting event of the next activity. However, one must always account for the natural constraints of the real world and provide a timing leeway cushion allowance as problems in actual production will require the ability of contingency planning for tasks and/or activities that may get bottlenecked along the way. Proper use of float is the tool is the method of contingency planning here which is known by the term *workaround.*

Event diagram. The most common event system uses circles at the ends of each activity arrow. These circles are placed at the junction of the arrows and they represent the event, or the moment of time at which an activity is just starting or finishing. Important events which mark the completion of a phase, such as a foundation final, are called milestone events for they serve as benchmarks to your schedule's progress versus run time.

Milestone. Date on the schedule predetermined for a phase or important occurrence is scheduled to take place.

Early start date. Earliest date an activity can start.

Late start date. Latest date an activity can start.

Early finish date. Earliest date an activity will be completed without float.

Late finish date. Latest date an activity will be completed without negative float.

I - J number. The letter *i* designated symbol for the tail of an arrow (start of an activity), and the letter *j* designated symbol of the head of an arrow (finish of the activity).

Float. Measure of available scheduling leeway time on any activity's completion.

Free float. Time by which activity finish event can be delayed without affecting the succeeding activity's start event.

Total float. Measure of available spare time or scheduling leeway available on all activities' completion. Total sum of all free float.

Negative float. Time a critical activity is late meeting its finish event.

Phases. Phases are groups of activities that will happen in a logical order, precede or succeed one another or happen simultaneously. As tasks are the micro units, phases are the project's macro units. Phases are arranged in divisions with sequential velocity management. Phases are supposed to flow into each other smoothly, on time, with no problems. If problems are anticipated, contingency planning in the form of workarounds need to be established ahead of time.

Constraint. A constraint is a potential real-world limitation, or a tactic of the scheduler in creating a float window during monthly recycling of the schedule, that can delay the starts of activities or tasks. Constraints, known as a dummy activity on network schedules, are shown as dashed lines with zero elapsed time. Constraints are negative factors that balance the calculations of job logic.

Job logic. This is sequential relationships between activities, identified and defined during preschedule planning. These relationships consist of the necessary time and sequential order of construction operations throughout the project.

Logic diagram. Arrow diagram of complete project network schedule, or a cross section of an area of production.

Time-Scaled chart. Logic diagram with a time scale.

Velocity diagram. Velocity diagram means a straight-line, time-scheduled flowchart. The purpose of the velocity diagram is to determine the most efficient paths to joining the activities in total operational network scope. Traditionally, the arrows in a velocity network diagram represent the activity itself, not the direction of movement. The angle of slope and arrow's length are not factors in the scheduling, simply the designer's choice. Each arrow in velocity network diagram represents an activity, identified by its activity number.

Velocity network schedule. This is the culmination stage of CPM. All the preceding scheduling elements are computed into a master network plan for the entire project's scheduling, including post-project closeout. Here, each activity is now also assigned its relative *i-j* number.

PERT. Program evaluation research technique. Mainly used by the military and defense industry contractors. Optimistic time = Earliest completion, shortest time. Pessimistic time = Latest completion, longest time. Realistic time = Normal completion, average time.

CPM schedule. A schedule using the critical path method of activities management.

PDM schedule. A schedule using the precedence (or node) diagramming method.

Developing the CPM Schedule

The first step in developing your schedule is to establish the "levels of specificity." It is important that you decide at the outset where you will put your planning energy. Consider which parts of the schedule and which aspects of the project are most important to your planning objectives. Set some general time expectations for each part of planning the schedule and try to keep them. This is a critical stage of the planning process, as you must answer each level of specificity question to avoid getting bogged down in the process or devoting an unreasonable amount of time to less important parts of the schedule. Consider the following in creating a level of specificity in developing your CPM schedule:

- **Sufficiency of past performance data.** How good is your company's historical information? Have the numbers shown a consistent pattern? If so, you can rely on them more; if not, find a better predictive source or database.

- **Study vulnerabilities.** Are there predictable risks? Do certain types of activities show poor event estimates in the past? How about the subcontractors? Are other construction projects similar to yours experiencing problems in areas of production?

- **Seek opportunities.** Are there particular areas of the project, such as atypical activities which would benefit from special attention or critical inspection before scheduling? Find the greatest opportunity areas and devote extra time to them. These are areas in which you can save the owner money and thereby improve your professional stature and reputation.

- **How much can you gain?** What is the potential for increasing production in each activity? If the potential is small do not spend much time on it, even though solving the problem might give you some personal satisfaction as a scheduler.

- **Time management.** How much time can you devote to planning and developing the schedule? What resources can be economically devoted to developing the schedule? This is often a tough balance. The answer here lies in what kind of business person you are. If you spend much of your business time in fighting business fires between contractors' work schedules in developing the schedule, planning is most important but least convenient. Consider your personal business tendencies: Are you generally obsessed with detail? Better then, in the interests of time management, to be a little more general to ensure getting the job done.

In summary, levels of specificity generally establish how much detail you will gather. You would be well-advised to do a detailed activity survey as part of your planning. You might also choose to research your competition via other project schedulers. What are they doing right? What are they doing wrong?

At this point, the development of the schedule depends on the project team's talking the project through. This is a critical part of the schedule's development, the project's management is now subject to careful scrutiny by all those who will have a hand in building it. Detailed examination of the project scheduling network logic comes next, by not only the owner and project team but by the subcontractors who will be doing the work items. The network diagram is then broken up into its phases, and the sequential order of the construction phases plus the timetables for those CPM phases are agreed upon or negotiated.

By the time a project owner or developer puts the project out to bid, a bulletproof CPM network schedule needs to be in place. Production scheduling usually starts right after the project's contract is awarded. At a precon (pre-construction conference), the general (or prime) contractor provides his overall timing schedule for the project, and then the activities subcontractors provide their scheduling within the parameters set by the prime contractor. Typically, if the owner has any experience in construction development, the owner or his agent, usually the project manager, will provide a contractual CPM schedule to which all three parties agree (or negotiate their positions). Which means your project schedule now becomes a contract document.

In a CPM schedule, arrow diagramming is most frequently used. The logic of an arrow diagram is graphically apparent to any reader. Obviously, the total project cannot be completed earlier than those portions of the work that require the *most* time to complete. This point of logic is what CPM is based upon. On complex schedules it is impossible to determine by examination of the network diagram which paths represent the critical path; thus computers have come into the picture. By using their speed and accuracy in computations, all possible combinations of tasks and activities times, the early and late starts and finish dates, and float times can be analyzed until all the key times can be determined from the computer printout. In addition, the computer determines the critical path, or paths, and shows any overruns in time along the critical path that will result in a total schedule overrun. This same program function can be set up for the schedule budget.

If performed manually, CPM does not offer the benefit of tabular printouts, which give key time data; thus the user must rely upon observation of the diagram itself. Although on very simple arrow diagrams, it may be possible to determine the critical path, it is normally necessary to construct a time-scaled network from the arrow diagram before any true scheduling can be determined. There are four basic steps in setting up a manual CPM network:

1. Establish the critical activities and their durations.
2. Determine the project's job logic and construct a dependency network.
3. Prepare a time-scaled diagram of the dependency network and determine critical path and float times.
4. Scheduler needs to obtain copies from all subcontractors of purchase orders of long-lead items to verify that the items have indeed been ordered and will be delivered on or before starting event deadlines.

There are also four important production timing facts that are determined from a CPM network:

1. Earliest time an activity can start and finish.
2. Latest time an activity can start and finish without completion delay.
3. The amount of float time available in that activity's scheduling.
4. How much of the project is now actually completed versus what is scheduled for completion.

These constraints are tools of time manipulation in CPM scheduling. For example, usually site development includes earthwork and underground excavation work that must be completed before foundation work can be begun. Conversely, the roughing-in of systems, such as fire protection and plumbing, can be performed at the same time, because they are similar in job logic but neither is dependent on the other. Activities subcontractors in these groups can finish work of simultaneous duration. Finally, the prime contractor estimates how long it will take the subcontractors to complete their activities. This estimate is made after the prime contractor consults with its subcontractors and reviews their schedules.

Total CPM management, therefore, can be seen as a structure made up of arrow diagrams reflecting constraints, precedence diagrams reflecting activity interdependence with predecessor and successor, and activities-on-nodes which show where the activities' events and durations fall. The CPM project scheduler begins by dividing the total project into different activities in different phases. Float time is used to act as an activity-linkage cushion for the network schedule, to handle the unforeseen variables and long-lead item delivery uncertainties that exist in construction projects. Next, the scheduler identifies the activities that must be completed before the other activities can be started and which will take the longest to complete, and a critical path for these activities. A commercial or industrial project may have thousands of activities, with just as many subcontractors, on multiple critical paths.

Figure 3-1 shows the production activities for a small job (garage), and their respective durations and constraints.

ACTIVITY	DURATION	CONSTRAINTS
Excavation	10 days	None
Formwork	7 days	Excavation
Concrete pour	4 days	Formwork/Plmbng
Plumbing	5 days	Concrete Pour
Electrical	4 days	Concrete Pour
Roof	6 days	All

Duration in Days

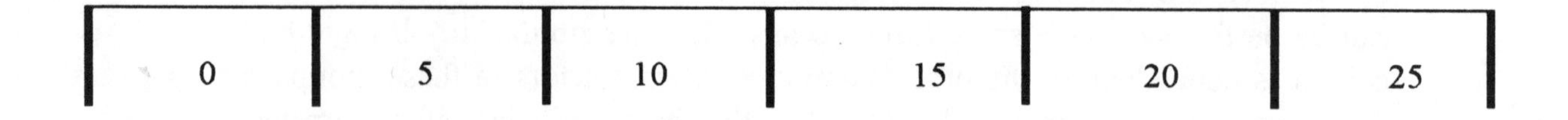

Constraints

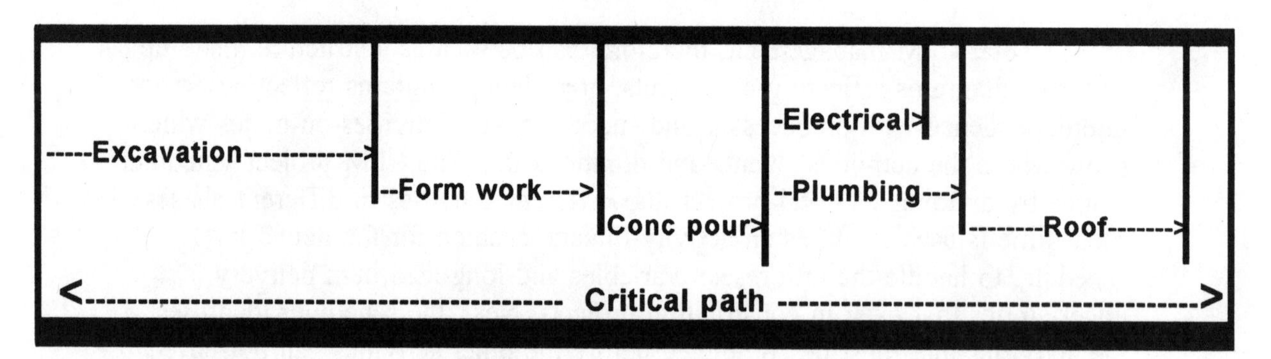

Figure 3-1
Activities, Durations, and Constraints

Constraints will control the critical path(s) of the CPM schedule. In Fig. 3-1, the excavation has no previous activity constraint so it can be performed first. Once it is completed, the formwork for the placement of concrete, which includes placement of steel, plumbing, and electrical, will be completed.

Now let's look at a larger residential project. The CPM schedule for this project appears in Fig. 3-2. The total project under this schedule should be completed in 90 working days. The critical path is shown as the longest path on this schedule, consisting of those critical activities that will cause a delay to the total project if they are delayed. In this example, all critical activities are on the critical path. A delay to any of these activities will delay the entire project. In contrast, lot grading and heating activities are not on the critical path. Their delay, up to a point, will not delay the total project. If the heating work is delayed five days, the total project will not be delayed. There is allowance for float in our example. The number of days each non-critical path activity can be delayed (free float) is:

Roofing...	... 3 days
Downspouts & Gutters...	... 5 days
Grade Lot...	... 4 days
Landscape...	... 5 days
Chimney...	... 5 days
Heating...	... 7 days

If any of the activities on the non-critical path (16-22) or (5-11) are delayed beyond their float period, they become part of the critical path. After the activity (3-4) Frame, the critical path is joined by five non-critical paths running in parallel. Some activities that were previously on the critical path will no longer be there. Suppose there is a three-day delay to (17-21) Grade Lot. Originally (17-21) Grade Lot had four days of free float. Now the CPM must be adjusted, because (17-21) Grade Lot now has only one day of free float. The total project has now been delayed one day of total float. (4-21) Heating has now become part of the critical path, and (16-17) Downspouts & Gutters has moved off the critical path.

Critical path activities, such as (2) Pour Footings, (4) Framing, and (11) Lath/Plaster Walls are not affected by changes in the deduction in free float of (17-21) Grade Lot.

Figure 3-3 shows the relative S-chart with a timeline for those activities.

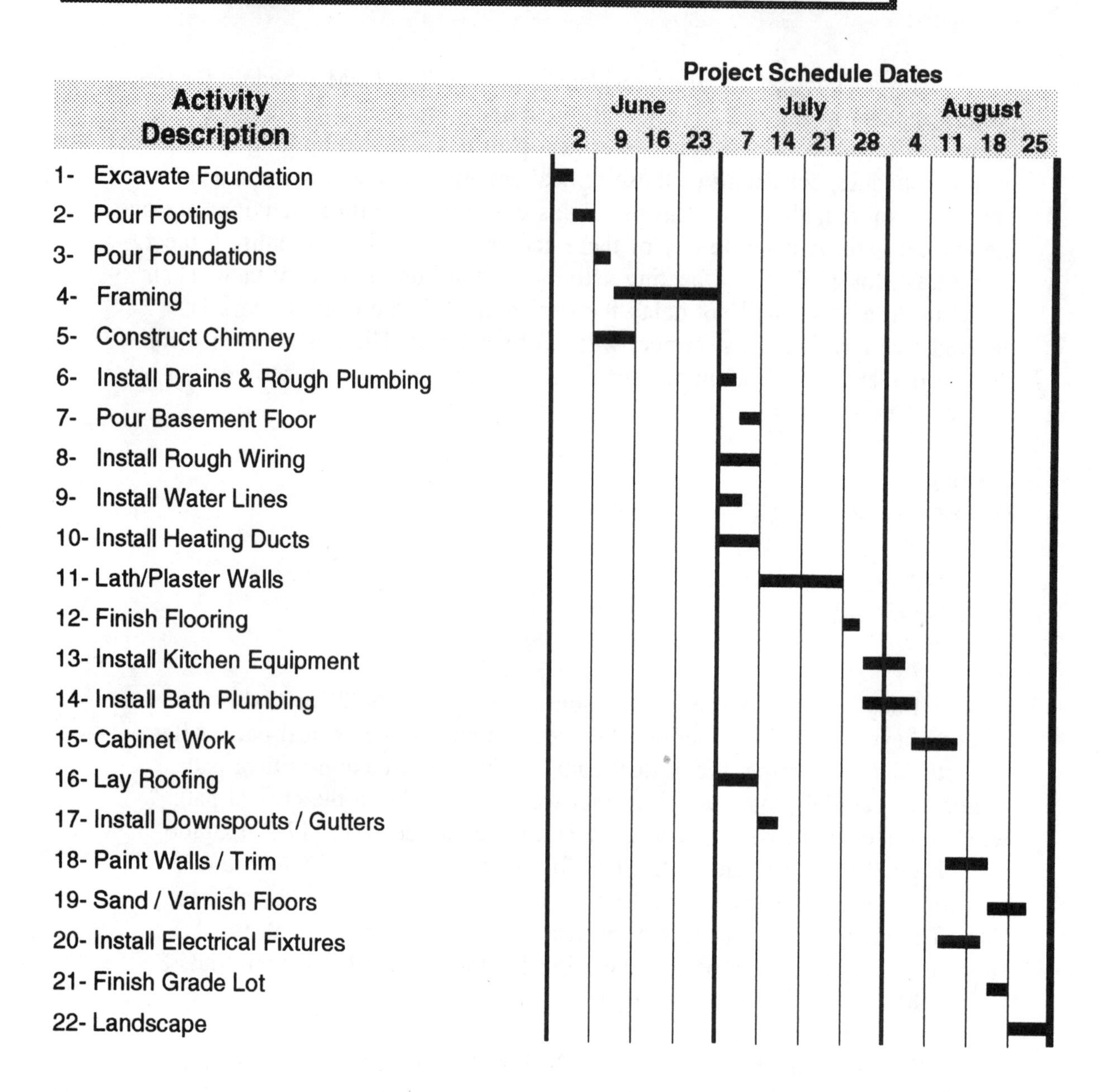

Figure 3-2
CPM Bar Chart

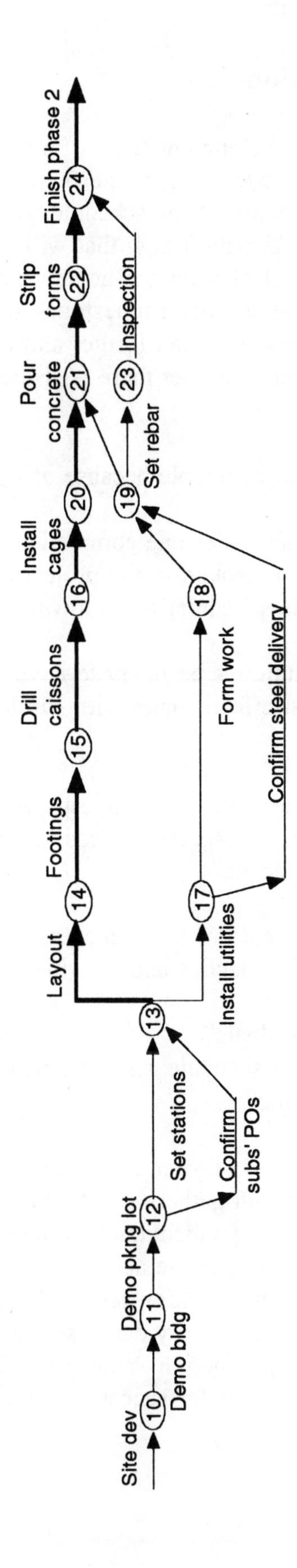

Figure 3-3
Schedule Network

Schedule Plan Evaluation

The last major step in the development stage of a CPM Network Schedule is evaluation. The evaluation process begins the final design of the schedule. Consider each activity and phase of the schedule. Also consider how you can begin gathering data and information now that will make future evaluations more effective. The first level of testing is achievable milestones matching projected budget costs. If the costs and milestones are right on target, is there any point in studying the schedule plan further and not just submitting it for approval and implementation? You bet there is! Here are some good questions that need to be answered:

1. Are the project milestones attainable because of our CPM scheduling effort or in spite of it?
2. Could we achieve the same milestone completion dates with fewer dollars spent? (This one question, spoken to the project team by the project scheduler will endear the scheduler to the owner forever. Don't fail to ask it.)
3. Will future budget expenditures be likely to have the same effect?
4. Could we repeat this cost-effectiveness with similar activities in other project schedules?

If the opposite is true (that is, the milestones and budget breakouts are not matching up), the reverse perspective needs to be applied. In such a case, these questions need to be faced:

1. Have we failed to implement our CPM successfully?
2. Were our time management tactics and activities objectives sufficient to attain scheduled milestone?
3. Were our milestones unrealistic?
4. Were there changes in our schedule assumptions during the plan period which affected its outcome?

To test the effectiveness of your CPM schedule during schedule recycling, it is essential to begin gathering data throughout the planning period. This may be as simple as asking subcontractors their opinions of previous jobs they've done, similar to their respective activity, keeping a record of their comments and averaging the information received. Or it could be as complex as a major push to obtain data from every database that has any conceivable bearing on similar activities production. Your database information networking efforts here will provide the basis for success in your schedule plan evaluation.

Performance Targets

Performance expectations are at the root of becoming and remaining a successful CPM scheduler. In the first instance, this is the process of setting production objectives: the percentage of activities completion per day and the relative cost-effectiveness projected versus actual cost. In each phase of production the prime contractor and activities subcontractors need to know exactly what is contractually expected of them.

In the second instance, performance targets are set to establish milestones for phases completion. Modern CPM management theory is evolving toward the position that contractors who will be affected by critical path milestone events should participate in creating those milestones during the schedule planning period. Accordingly, as part of the planning process, current CPM standards and methodology should be reviewed periodically before the schedule is implemented. Performance targets should be concentrated around areas of progress development or production constraint, which should then be critically examined and options or workarounds considered.

Project Planning

Advance planning of job logic prior to developing the phases of scheduling is a crucial CPM primary step. The importance of planning is often overlooked in the anticipation and excitement of doing the deed. I could go into all the reasons I've encountered over the years about the importance of advance planning, but an author who lived before my time put it much better:

> He who every morning plans the transactions of the day, and follows that plan, carries a thread through the complexities of the most busy day. The orderly arrangement of his time is like a ray of light which darts itself through all his occupations. But where no plan is laid, where the disposal of time is surrendered merely to the change of incidents, all things lie together in one chaos.
>
> Victor Hugo

I doubt if Victor ever did CPM project scheduling but his words ring true today just as they did a hundred years ago. Chaos is the anchor that must be cut loose for a CPM schedule to perform properly. Prior planning prevents chaos and the variables that cause chaos, thereby reducing risk. The planning of any network schedule must start from a work breakdown structure (which is the detailed activity list) showing each activity to be scheduled. Sometimes it looks like an activity tree, as one major item is subdivided into its major parts. For example, we might have an activity called "install foundation," which could be further subdivided into "excavation," "forming," "setting rebar," and "pouring concrete." The foundation activity could be further subdivided into "buildings" (affecting structural steel erection) and "equipment" (affecting equipment delivery and erection). The decisions affecting the degree of work activity breakdown will determine the length of the activity list and the complexity of the overall construction schedule. As everything in CPM is series-strung, one thing leads to another. All decisions affect the whole. Conversely, the greater the detail the greater the control.

Accordingly, the first step in CPM scheduling is the planning stage. The project must first be broken down into its activities and related tasks. Activities and tasks are the two basic components of CPM. The extent to which the project is subdivided into tasks and activities depends upon the size of the project; however, the project's sequential planning steps are always the same:

1. **Create milestones.** Planning establishes expectations and goals for your project schedule. It requires project-wide coordination and project team commitment to common goals of achievements in critical activities.

2. **Establish priorities.** Planning requires you to consider basic goals, policies, plans, and resources, and their relative importance. By establishing priorities, you also establish a relative scale of importance to assess your resources allotments.

3. **Predict problems.** Planning predicts problems and production bottlenecks in activities and operations. It invites the creation of controls and evaluation procedures.

4. **Solve problems.** Contingency planning allows you to evaluate what doesn't work, and then to consider what might be done to unscrew the situation. With a plan detailing your assumptions and expectations, it is much easier to identify what went wrong when problems arise in the project schedule. Even guessing will allow you to check and improve areas of concern within the schedule.

5. **Go for profitability.** Planning forces consideration of the financial issues and owner's cost before resources are committed. It imposes a test on intuitive choices, points out options, and allows for you to consider alternatives.

6. **Establish motivation.** Planning also encourages full participation from the project team and motivates everyone concerned, including you, to put forth their best efforts.

7. **Create cost control.** Planning controls costs by creating a demand for periodic reporting, audit tracking, similar comparisons, and cost-effectiveness evaluation. This is a big area of concern to all owners and is one of the major advantages having a professional project scheduler on the project team, as architects and design professionals traditionally have shown a casual attitude towards owners' cost control.

8. **Encourage development.** Planning is the only way to put milestones into the operational phases of the schedule accurately. It requires considered decisions, rather than default outcomes, balancing needs and demands for resources.

If we boiled all this down into short little title tags to aid in memorizing, the list of item name tags would look like this:

- Establish objectives
  - Organize
    - Develop project team
      - Motivate
        - Hire contractors
          - Communicate
            - Measure
              - Record
                - Analyze

If these steps are considered activities, some of the relative tasks within each activity would be steps like these:

1. Designate different areas of activity completion responsibility, such as subcontracted work, that are separate from that work being done directly by the prime contractor.
2. Spreadsheet list of different categories of work as distinguished by trade.
3. Spreadsheet list of different categories of work as distinguished by equipment.
4. Spreadsheet list of different categories of work as distinguished by materials.
5. Breakout distinct and identifiable subdivisions of structural work.
6. Locate all work within the project that necessitates different timing.

7. Identify all long-lead items that will be special ordered.
8. Use owner's breakdown for bidding or payment purposes.
9. Use contractor's breakdown for estimating or cost accounting purposes.
10. Use weekly summary reports to track progress.

The activities chosen may represent relatively large segments of the project or may be limited to small steps only. For example, a concrete slab may be a single activity or it may be broken into separate steps necessary to construct it, such as erection of forms, placing of steel, placing of concrete, finishing, curing, and stripping of forms or headers. As the separate activities are identified and defined, the sequential relationships between them must be determined. These relationships are referred to as job logic and consist of the necessary time and order of construction operations. To develop a project schedule, one applies these crucial questions to the preceding steps:

- What is the estimated time duration of this project? Exact start date? Finish date?
- How detailed a schedule is appropriate? How many reports or sorts do I need?
- How often will I update this schedule?
- Who needs to receive information about specific areas of progress? When?
- What kinds of field reports will I need? From whom and when?
- What computer resources will help me schedule tasks best?
- How much time can I afford to spend on this specific project schedule?

Now "flesh out" the schedule by adding these details:

- Determine in your mind how to effectively build the project.
- Identify each task necessary to complete the job, and make a detailed list.
- Determine how these tasks will group together to form activities, and make a detailed list.
- Estimate how much time will be needed for each activity.
- Define how the activities relate to each other. Build a network task flow diagram that clearly shows relationships between activities.
- Assign a responsible subcontractor to each activity so that, when you update the schedule, you'll know who's responsible for each phase.
- Create an interactive activity flow diagram with the contractors and subcontractors who will do the work.

- Modify the task flow diagram until activities seem to flow correctly.
- Identify the critical path of production. Remember, CPM in its simplest form means establishing the chain of activities that will require the most time to complete.
- Seek ways to simplify the project. Explore any options to compress the schedule by performing activities in parallel.
- Consider whether you will have sufficient resources to accomplish several tasks at once.
- Eliminate negative float by modifying the network. (We will further define float in Chapter 5, and will examine its positive effects in Chapter 6 and its negative effects in Chapter 8.)
- Identify all long-lead items. Long-lead items are those materials and items which are not readily available on the local market and must be ordered, fabricated, or manufactured. Refer to Chapter 13 regarding back-up Purchase Orders for subcontractors' long-lead items.
- When satisfied with this basic schedule, apply the resources to the activities to build a complete schedule. Although the schedule indicates required actions and when they must be done, resources such as people, equipment, material and money actually do the work.
- Double-check to make sure you will have the required resources when you need them. Failure to do this step always results in cost overruns and will undoubtedly cost you your job.
- Juggle the schedule to resolve conflicts with other activities that use the same resources, and reallocate resources if necessary. Use constraints in reallocating resources, because all activities are interdependent and any juggling on your part will have long-reaching effects.
- Check that your schedule doesn't call for more than the normal availability of resources. This is one of the pitfalls the inexperienced project scheduler often encounters.
- Graph out and level your resource plan. Examine the resource-use profiles to determine whether the schedule contains hard-to-manage peaks and valleys.
- Consider the combination of time and money your schedule represents. Could you deliver the finished product sooner if you had more money or resources? Are these factors worth thinking about before you seek approval of the schedule? Compare costs, list requirements, check assumptions with the project owner(s), and refine the plan.
- Organize your project's schedule information by categorizing activities by phase, responsibility, department, and location.
- Set priorities and progress milestones along the schedule.
- Make your computer summary reports easy to retrieve and interlink the files to automatically analyze field progress reports. Summarize all details.

- Be sure to regularly record the status of every activity. Keep track of who did what and how close they came to projected scheduling so that you can improve future schedules. Gather progress reports regularly, preferably daily. Record how long it takes to perform each activity, what percentage of the activity is actually accomplished, and how much more time is required to finish the activity. Make sure the data you use for your analysis are accurate. Any data processing system, such as a CPM network schedule, is only as good as the data and information entered into the system for computation.

When the time sequence of activities is being considered, constraints must also be considered. Constraints are the practical limitations that can influence the start of certain activities, and are shown on CPM schedules as dummy arrows. For example, an activity that involves the placing of reinforcing steel obviously cannot start until the steel is on the site. Therefore, the start of the activity of placing reinforcing steel is constrained by the time required to prepare and approve the necessary shop drawings, fabricate the steel, and deliver it to the job site. CPM project schedulers will timeline constraints just the same as activities, and will display them as dummy arrows with durations on the network diagram.

A comparison of the amount of detail covered in a CPM work breakdown structure, and a comparable bar chart, clearly shows the difference in the amount of detail covered by the two systems. Bar charts cannot even approach CPM schedules in activity numbers without becoming completely unworkable. Another outstanding advantage of CPM is the intangible benefit of forcing the project team during the planning stage to dissect the project into all of its working parts. This forces the early critical analysis of each work activity. The project team will then run the first pass of the schedule several times, to test and debug the logic diagram before the final version is ready for review and approval. So your initial CPM logic diagram equates to a practice run on computer for the full-blown project schedule.

Experienced owners and developers are aware that the actual scheduling phase, such as calculating the early and late start dates with interlinked free float, is best left to the professional project schedulers who can cook the schedule into that final, debugged, logic diagram. They will want to see a smoothly flowing job logic within the network schedule that they can relate to as being realistic and achievable. Contingency planning here by the scheduler, which provides alternate critical paths, is looked on favorably by the owner and project team. Another hot spot to take note of.

Scheduling Budget

A scheduling budget allows you to keep control over cost expenses. It shows you what you need to do to make the schedule operate successfully. It indicates where resource allocations have priorities, and in this sense a scheduling budget is used to project goals to work toward. Meeting these goals means expenses should match projections by the end of each month. And if they do not, you should immediately adjust the scheduling budget, based on the newly revised anticipated cost expenses.

Budget costs in construction projects are divided into two major categories: direct costs and indirect costs. Direct costs are those expenses that are specifically by the project, such as labor and materials. Indirect costs are those expenses which are the costs of doing business not directly related to any one job, such as operating expenses and overhead. If the indirect cost budget does not allow for the expense of a sophisticated scheduling method, and the cost of training personnel required to use it, the project is going to come up short of money. Most computerized CPM scheduling costs have a tendency to grow and overrun their budgets because of just such an oversight.

A common problem is job stretch-out, which increases the schedule cycles, which in turn runs up the scheduling personnel hours and computer time. A factual estimate of the total cost of the proposed scheduling system is needed if an effective system for the project is to be selected. A second common problem I have often encountered is mixing cost control and schedule control. This, in my experience, has never been very successful on larger projects. The problem here lies in the fact that the budget numbers are broken down into different categories than the schedule activities, and cost-tracking systems need the numbers of both the budget and the schedule to be separated for cost-accounting accuracy in the audit trail.

Combined schedule and cost-reporting systems have not yet become sophisticated to the point of overcoming intermittent data entry such as this, and accordingly cannot generate audit-trail accuracy. Although these scheduling systems are cheaper to purchase and operate, my advice, based on experience and statistics, is to use caution when employing a system that ties those two key areas of project control together, and thereby runs the risk of losing control of two very critical areas of project management. This is especially true of larger projects with multiple critical paths. Schedule control and cost control need separation for individual control.

In order to make the most efficient use of remaining resources in the scheduling budget, the scheduler must look carefully for cost overruns. Checking for cost overruns means determining where cost exceeded expenses.

This is money that could have been applied in the project in a more efficient manner, is now lost, and must be made up by shaving money from succeeding activities or from owner's profit.

Human Resources Leveling

Another major advantage of the computerized CPM system is the ability to level out labor resources peaks during the project schedule, in workforce labor and administrative personnel requirements. These peaks happen intermittently throughout the project's design and construction phases. By taking advantage of the available float and rescheduling the start of noncritical activities, it's possible to shave some labor force and personnel peaks. Leveling the personnel requirements leads to more cost-effective use of the project's human resources by adjusting the costs to phases, and deleting duplication of effort. This option is invaluable for smoothing out the different work trades manpower peaks in key areas of the work. Judicious use of the early and late start dates can also keep subcontractors from getting in each other's way, which results in delay for those interrelated activities.

Selecting specific areas of human resources leveling allows the scheduler to level only a select area, or range, of the network schedule. This select critical analysis cuts down the time needed for leveling operations. It also improves the computer's variables needed by the program to calculate "what-if" analyses. As critical activities are completed, the network's human resources leveling decreases as activities are omitted. And as the human resources needs will vary from project to project, the scheduler needs to utilize the select areas of leveling that are applicable to the specific circumstances.

An important aspect of human resources leveling is that key project team personnel should be thoroughly trained in CPM. That includes all levels of personnel from the design architects through the construction management team, and should also apply to field operations supervisory personnel. It is not recommended that a newly trained crew be controlling a large project, nor using a new software system without running the old scheduling system in parallel, at least until the new system has proven itself to work and the personnel running the system have become familiar with it. If the new system breaks down for any reason, and you don't have a back-up system that's been running in parallel, you will be without any way to control the completion event on the project, and on a critical path to litigation.

Matrix Networking

The most sophisticated (and currently emerging) CPM systems incorporate "matrix networking," in which specialized subcontractors and activities are utilized in a variety of project-oriented configurations. The project manager works with small teams from the various areas of specialization in the project, toward common ends of interlinking completion events.

This is a highly responsive and event-oriented operations system, producing a matrix network that will suffice as divisional factors in the CPM schedule, thereby reducing time required to manage multiple schedules. This is one of the basic elements in fast-tracking. However, matrix networking requires sophisticated subcontractors and project team members with good communication skills and a commitment to the project's schedule.

Matrix networking is a process of determining:

- The tasks and activities to be performed
- The time, in events, to be devoted to each
- The job logic sequence of each activity
- An optimistic and pessimistic completion timeframe

By combining these four tasks into a timeline matrix we cut down on some of the data processing time on most of the schedules, if we are running multiple projects and multiple schedules. And, if those schedules are similar. Controlling the CPM network schedule by matrix networking is a combination of planning and presetting policies for operations that include activity milestone summary sorts. The advantage to using matrix networking is that if you do so, the summary sorts become evaluation mechanisms *before the fact*. This gives the owner an edge just as fast-tracking does.

This type of controlling is an area of opportunity for most small- and medium-sized projects running budgets of five to ten million dollars (that is a medium-sized construction project in California, according to SBA statistics). The development of cost-efficient processes for assessing performance expectations, measured by milestone indications of success, coupled with remote (field) data collection systems linked to a host processing computer back at the office, is fast becoming the difference between success and failure in CPM scheduling.

As the schedule grows in complexity, the functions of controlling CPM through matrix networking also become increasingly complicated. At this point, it is critical to keep in mind that it is not necessary for you to be a statistician to begin analyses of these time-manipulation factors. Remember, that's what the computer is for. Set up the spreadsheet formulas to do the range statistics calculations for you. Spread your databases over many ranges and remember to leave a blank row or column between each database for additional safety in data retrieval. The big picture is what's important, and where you need to stay focused. So the more you understand about the limits of your schedule and the elements which define as well as constrain the project's production events, the better you will be able to manage time manipulation of those events.

Chapter

4

Network Scheduling

Activity Arrows

In network diagramming, the symbol for an activity is an arrow. The arrow represents linear timeline movement from left to right, start to finish. The production continuity is the arrow moving from a preceding activity to a succeeding activity. Arrows in a velocity network diagram represent the activity itself, not the direction of movement. The angle of slope or the arrow's length are not factors in the scheduling, simply the designer's choice. Each arrow in the velocity network diagram represents either an independent activity or an interdependent activity. Each arrow is identified by its respective activity numbers.

In the example diagram in Fig. 4-1, the activity arrow designated (3-7) is "install caissons." The according tasks that will fulfill that activity, in their order of precedence, are:

1. Accept winning bid and award contract.
2. Schedule municipal inspection.
3. Order and deliver reinforcing steel cages.
4. Layout the exact location of the hole.
5. Excavate caisson.
6. Install rebar cages.
7. Connect cages to foundation continuous rebar.
8. Place concrete.
9. Strip forms.
10. Clean up site.

In Fig. 4-1 we see the sequential path that represents the job logic of the production. Although the activity (3-7) Install Caissons is in actuality the task of pouring concrete into the predrilled holes over the preinstalled rebar, the task is dependent on activities 1 through 7 having been completed before the concrete truck shows up on the job site. If any of those activities are incomplete or delayed there will be a direct effect on activity 8.

Thus we have established the critical path for this part of the network schedule.

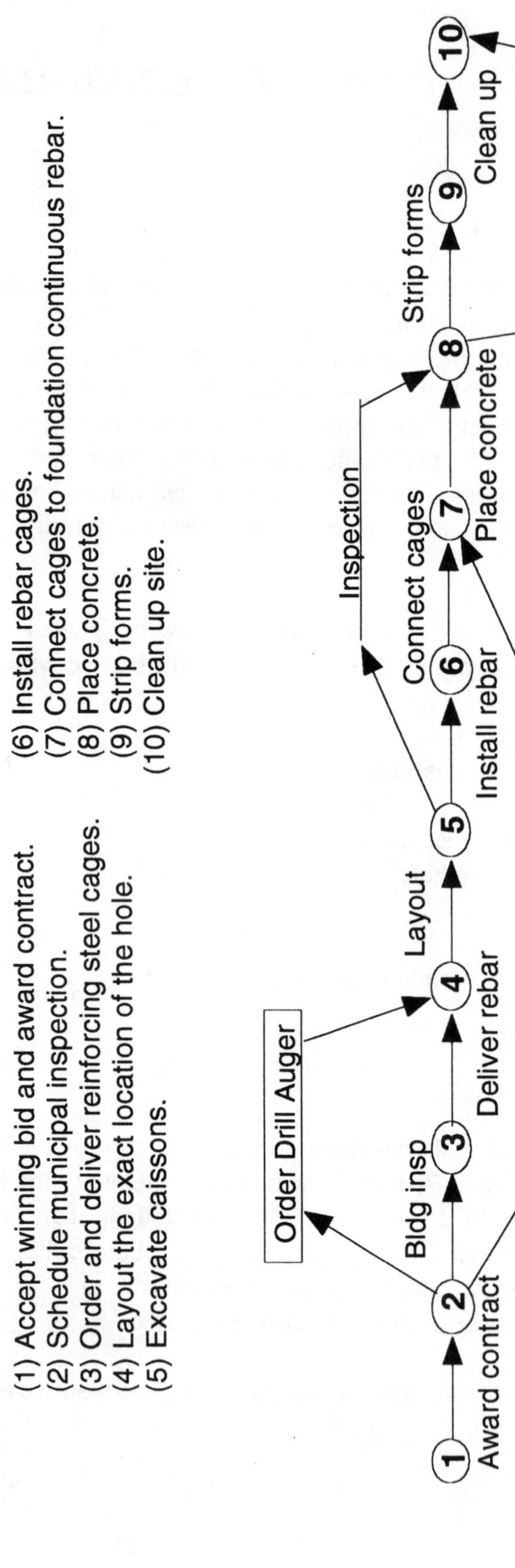
(1) Accept winning bid and award contract.
(2) Schedule municipal inspection.
(3) Order and deliver reinforcing steel cages.
(4) Layout the exact location of the hole.
(5) Excavate caissons.
(6) Install rebar cages.
(7) Connect cages to foundation continuous rebar.
(8) Place concrete.
(9) Strip forms.
(10) Clean up site.
Order Drill Auger
Inspection
Award contract
Bldg insp
Deliver rebar
Layout
Install rebar
Connect cages
Place concrete
Strip forms
Clean up
Order and deliver concrete
Phase 1 Closeout
1
2
3
4
5
6
7
8
9
10

Figure 4-1
Job Logic

Dummy Arrows

Dashed-line arrows in CPM diagramming are the symbols for dummy arrows, which are diagramming symbolism showing the constraint dependency between activities. Any dummy activity acting as a constraint is shown as a dashed line with zero elapsed time. The direction of the arrow shows the production order of activities and a dummy arrow shows what activities constrain the start of another activity or activities.

In Fig. 4-2, the dashed arrow (7-8) is a dummy arrow showing the constraint of (8-12), starting until both (6-7) and (6-8) have finished. There is no similar constraint upon activity (7-12), so it can start when (6-7) finishes. It is not dependent (constrained) on the completion of (6-8). It also shows that the CPM job logic must move forward from this activity. If two or more activities begin to run simultaneously, the computer reads only the same *i-j* number of both activities. So the project scheduler gives each activity its own *i-j* number for the computer to use as a relative cell address for that specific data, showing it on the network diagram as a dummy arrow.

In Fig. 4-3, five activities in this network diagram can start and finish at the same time, and be completed in timelines parallel to one another. However, the computer reads only the same *i-j* number of (21-28) for all five activities, thereby making it impossible to separate the activities in the computer program. To overcome this lack of separation and control, the project scheduler uses dummy arrows to assign *i-j* numbers to each activity, thereby giving each activity its own computer cell address. Dummy arrows are also used by the project scheduler to assign different *i-j* numbers to any activities that are running in parallel.

Figure 4-3 shows how these five parallel activities can be run simultaneously and can be further utilized in time management as parallel activities adding to total float. By the use of distinct *i-j* numbers assigned to each of the five activities, dummy arrows are used as shown in Fig. 4-4 to show the interdependent relationships. Now the five parallel activities shown in Fig. 4-3 retain the interdependency relationship of each activity to the other, and are further identified by distinct *i-j* numbers. The computer program now identities them as activities (21-24), (21-22), (21-23) and (21-25), as seen in Fig. 4-4.

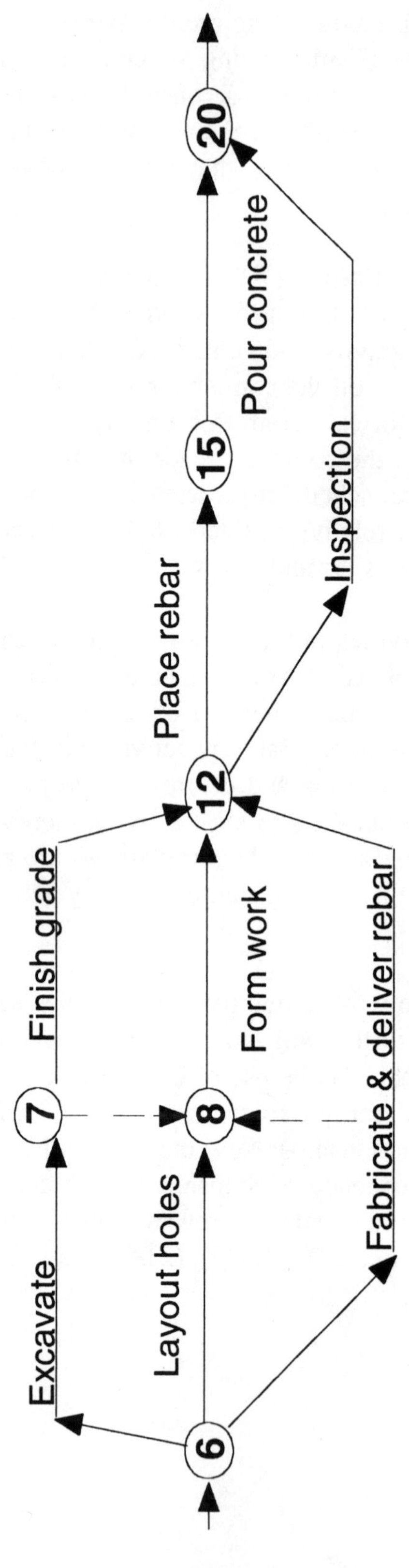

Figure 4-2
Dummy Arrows

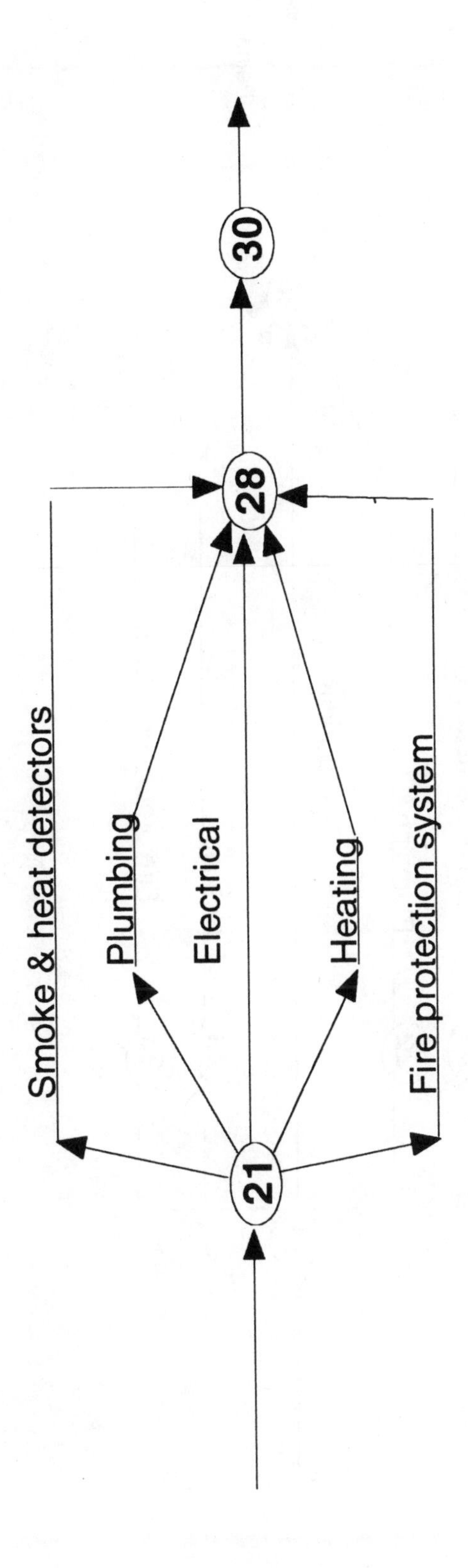

Figure 4-3
Parallel Activities

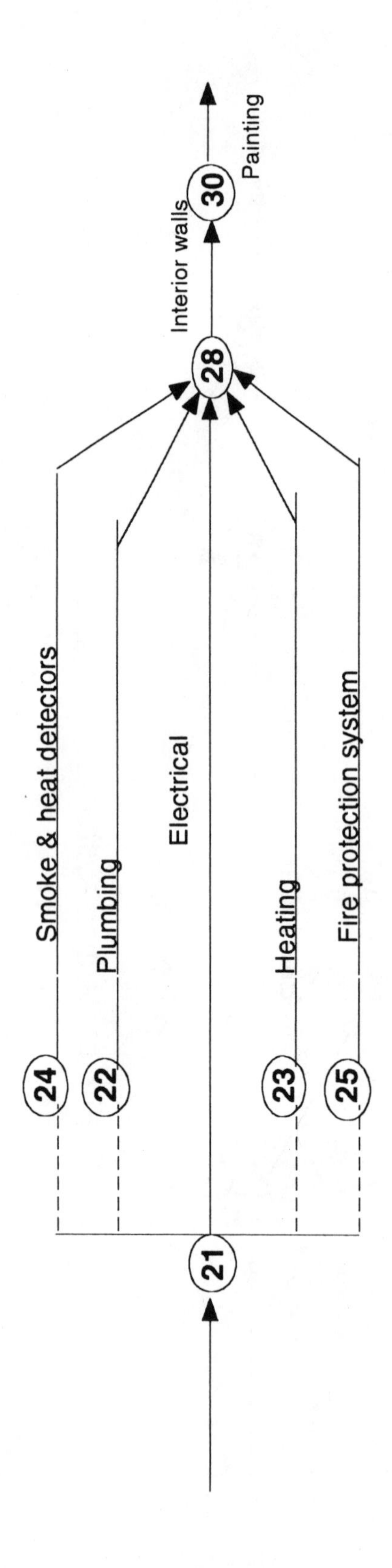

Figure 4-4
Parallel Activities Constrained by Dummy Arrows

I-J Numbers

In CPM, events are the exact day an activity starts or finishes. They are also dates of milestone completions. Events are assigned identification numbers for computer processing. The starting event number is the *i* number, and the completion event is the *j* number. The *i-j* number is used as a relative cell address for recording the activity's data. If CPM diagrams were to be prepared using random activity numbering, all activity numbers in the entire network would have to be renumbered to allow any new or changed activity to fit in a sequence with the other activities. Therefore it is a basic CPM requirement that, when event numbers are assigned, the finishing event number at the head of the arrow must be greater that the starting event number at the tail of the arrow, and the *j* value of each activity must be greater than its *i* value. A typical CPM network can involve hundreds of separate activities that must remain flexible in scheduling, so the experienced project scheduler assigns the activities *i-j* numbers only *after* the entire network has been completed and is ready for its first cooking or trial-run computation.

Using the vertical and horizontal axes of graph coordinates, , i-j events can be displayed in either plane. The vertical numbering method is more widely used, which numbers all events in a vertical column in sequence from top to bottom that equates to a parallel timeline, those groups of activities then moving from left to right. There is no significance to the event numbers themselves except as means of identifying activities, so if the CPM format of keeping the *j* value of each activity greater than its *i* value is used, blank cells can be left in the numbering system so that spare numbers are available for changes or additional work that may come up. Sequential *i-j* numbering provides this flexibility in scheduling, while also providing the computer with program logic data for events and activities locations on the network diagram.

Milestones

Milestones are benchmarks of long-term and short-term progress events that require some reasonable progress in activities or tasks toward a desirable outcome. Milestones are specific, time-limited objectives which, in combination, are sufficient to achieve those progress goals. Milestones are placed in the schedule to produce:

- Motivation for project team and contractors
- Utilization of competitive advantages
- Repair schedule competitive disadvantages
- Acceptable progress to the owner and project team if those milestones are

are met on time and within budget

- Measurement of actual progress versus scheduled progress

At this stage of the process in installing milestones, the scheduler should be thinking in broad terms. As one studies particular areas of the schedule, one develops progress objectives that can be related as milestones. Milestones should reflect both strategic planning as well as current period objectives. They should also represent success, not the absence of failure or the remediation of existing problems. This is an important concept for the professional project scheduler to grasp and use in developing the schedule. Unscrewing a problem isn't progress if the problem should not have happened, so milestones should mark progress only. Positive benchmarking is important; milestones should be specific and measurable objectives. They should also appear as projections of progress evaluations at the end of each construction phase. Interim milestones should measure phases of progress which are critical to meeting long-term project timing. They should "back down" from the specified completion date to the present in sequential order.

Once you have identified milestone phase areas, schedule dates for those events and the completion objectives that go with them. If you are scheduling with a group, such as the project team, it is important to agree upon the goals and objectives of each phase, leading to the establishment of milestones as benchmarks to evaluate the progress towards those goals and objectives. It is necessary to agree upon milestones prior to establishing phases, because milestones should lead the completion of each phase.

To reach CPM milestone events on time it is necessary to back down to the present. If you are to reach the critical completion date, where must the project be in one month? In two months? Next quarter? To actually achieve your CPM milestones it will be necessary to periodically re-evaluate start and finish events and use float to close gaps in timing. In planning the schedule, don't spend unreasonable amounts of time on making "perfect phases," but instead create obtainable and realistic milestones by which you can measure the project's progress.

The best planning process is to estimate durations based on what you know, and your company's historical data. Include in the schedule periodic sorts, or summary reports, that will test the milestones objectives for the coming quarter. This tactic will improve future milestone projections. Remember, the purpose of planning is to reduce risk. The best schedule plan is one that allows you to test your estimates before they cost lots of money. This is always your client's perspective and should be yours as well. We schedulers do this by running "What if?" scenarios with varying activity early and late starts, and varying the activity early and late finishes. A test of the schedule's quality is the extent to which completion events are balanced against float. Every project schedule needs to be built upon this basic criteria:

1. Each macroactivity step, defined into phases. This should be detailed enough to be measurable and give guidance to those who are responsible for managing that area of work.

2. The person responsible for seeing that the activity is completed. This is usually the prime contractor, the site supervisor, a project team member, or an activity subcontractor.

3. Activity cost in relationship to total budget. This should include time estimates, indirect, and direct costs. In the budget sort, a calculation needs to appear with this data to build a cost-tracking analysis.

4. Project phases deadlines. These should include the actual time available by the subcontractor responsible for the activity, including its assigned early and late start and finish events.

All of these four steps should be closely monitored up to and including project closeout. In larger and more sophisticated projects this basic outline of plan elements will be more complex, but still follow these basics in a larger configuration. Time and cost estimates are made in greater detail for each activity. "Event dependencies" are established, which must be accomplished before others can be begun. Resources that must be devoted to each are listed, then broken out from the total budget. Resources are then allocated with priority weighing, to ensure critical path activities are resource-covered first. Progress-checking milestones and specific sorts are built-in to the procedure to ensure project schedule compliance.

Job Logic

The logical sequence of the project's construction activities, factored by local practical limitations, is referred to as job logic. The activities chosen may represent relatively large segments of the project or may be limited to only small steps. To use a previous example, a concrete slab may be a single activity on a small job, but on a larger job it will be broken into the separate steps necessary to construct it, such as excavation, sub-ex preparation, erection of forms, placing of steel, placing of concrete, finishing, curing, and stripping of forms. As the separate activities are identified and defined, the sequence relationships between them must be determined. These relationships are referred to as job logic and consist of the necessary time durations and sequential order of typical local construction operations that are unique to your geographic area.

It is a basic fundamental in CPM that each activity must have a determined starting event, which may be either its own start or the finish of the preceding activity. Activity durations cannot overlap their finish events. Therefore, job logic is established to provide operations sequences within practical constraints.

Established job logic is then used to build program logic within the computerized CPM program. By determining the job logic, activities can have their interdependencies critically examined during all phases of the schedule, before errors occur costing delays and money.

Logic Loops

The logic loop is a paradox in network planning. It indicates that a critical activity must be followed by another critical activity which has already been completed. Even as I read that last sentence I realize it doesn't make any sense, and I'm the one who wrote it. But bear with me and I'll try to explain. The term logic loop is really an oxymoron, since a logic loop is anything but logical. Logic loops should be called illogic loops, but again, we professionals like to keep our industry jargon as confusing as possible. If, in our network schedule, one arrow was inadvertently connected to the wrong node, its path might well be shown to the computer to run backward. Although this seems obvious in a simple, single-path CPM schedule, if we multiplied the paths by hundreds of activities such as in large commercial or industrial network diagrams, logic loops can be easily overlooked in the planning stage.

In the vertical method of notating event nodes, which is more widely used, numbers of all events are in a vertical column in sequence from top to bottom, which equates to a parallel timeline. Because of the vertical configuration, activity job logic can have logic loops in those vertical groups of activities without the scheduler realizing that they are there, the error then moving from left to right on the timeline with the activity group.

I advise you to study the network diagram carefully at the beginning of the development of the schedule to confirm the job logic of the structure and to search carefully for logic loops. The best method to safeguard against them is to use sequential *i-j* numbering. The computer sort printout cannot indicate any clues to the presence of logic loops under the use of a random *i-j* numbering system, so random numbering guarantees a greater likelihood of error by allowing logical loops to remain undiscovered. Accordingly, to make ultimate use of computer program logic, the database, which in this case is the *i-j* numbers, we must number the activities sequentially.

Program Logic

Program logic can best be described as the way the computer puts together the time sequence for the activities involved in the project. The program logic for the sequence of operations in our example of Fig. 4-5, "Installing Caissons," would be based on the following job logic:

Activity	Sequence	Symbol	I-J #
Accept winning bid and award contract.	1	BC	1
Schedule municipal inspection.	2	MI	2
Order and deliver reinforcing steel cages.	2	OD	3
Layout the exact location of the hole.	2	LC	4
Excavate caisson.	3	EC	5
Install rebar cages.	4	IR	6
Connect cages to foundation rebar.	4	CR	7
Place concrete.	5	PC	8
Strip forms.	6	SF	9
Clean up site.	7	CU	10

This is the type of data strategy that is decided by the project scheduler initially, then by approval of the project team in the planning stage. For the purposes of CPM, job logic requires that each of the activities in the network have a definite event to mark its starting point, and another to mark its completion point. This event may be either the start of the project or the completion of preceding activities.

It is a basic tenet of CPM that the finish of one critical activity cannot overlap the start of a succeeding critical activity. When this happens, the work must be further subdivided into more detail. It is a fundamental rule of CPM that a critical activity cannot start until all those critical activities preceding it have been completed.

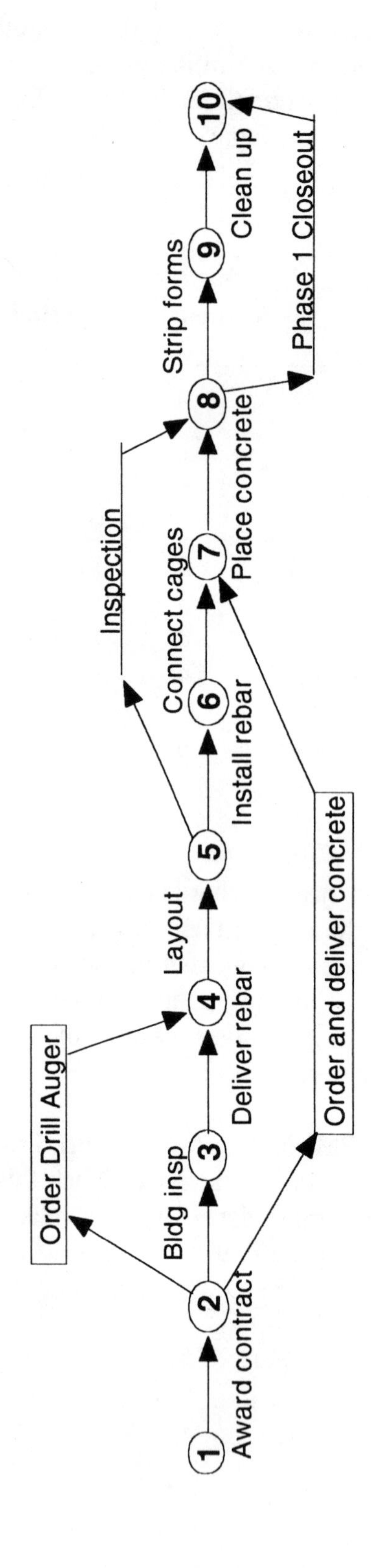

Figure 4-5
Job Logic Factoring Program Logic

Logic-Based Scheduling

Now we'll take all of the previous elements and combine them into an example that will illustrate logic-based scheduling. Examine the basic layout of a small commercial project's logic-based schedule, as shown in Fig. 4-6. Because all phases have similar traits but different activities, we will examine one phase of the schedule only. The foundation's phase activities, in their job logic, include all subactivities up to and including installing the building's foundation.

11. Demolish existing structure.
12. Remove existing parking lot.
13. Survey and set engineering stations.
14. Layout footings, caissons, building envelope.
15. Dig footings.
16. Drill caissons.
17. Install utilities.
18. Form work.
19. Set rebar.
20. Install cages.
21. Pour concrete.
22. Strip forms.

The foundation phase begins as soon as the site development phase finishes with the demolition activity, which can be done as the last part of the site development or the first part of the foundation clearing, thereby having free float that the scheduler can use to either constrain or accelerate the activity event. The phase begins with demolition of the existing building and the removal of the old parking lot. These events run in series because they can be done by the same subcontractor simultaneously. Activities (17) and (14) can run in parallel, as layout can be done right behind the surveyors. Activities (15) and (18) run in parallel, because they can be done by the same subcontractor at the same time. However, activities (14) through (21) are in series because each requires the completion of the preceding activity.

Each activity has subtasks within it that require sequential completion. Subtasks within (19) Set rebar, for example, include order and fabrication of the specified reinforcement steel, delivery, and placement of that steel. (20) Install cages would also have a final subtask of pre-pour rebar placement inspection by the municipal inspector before (21) Pour concrete could be done (usually the next day). Each activity's duration is given in events and timescaled in working days.

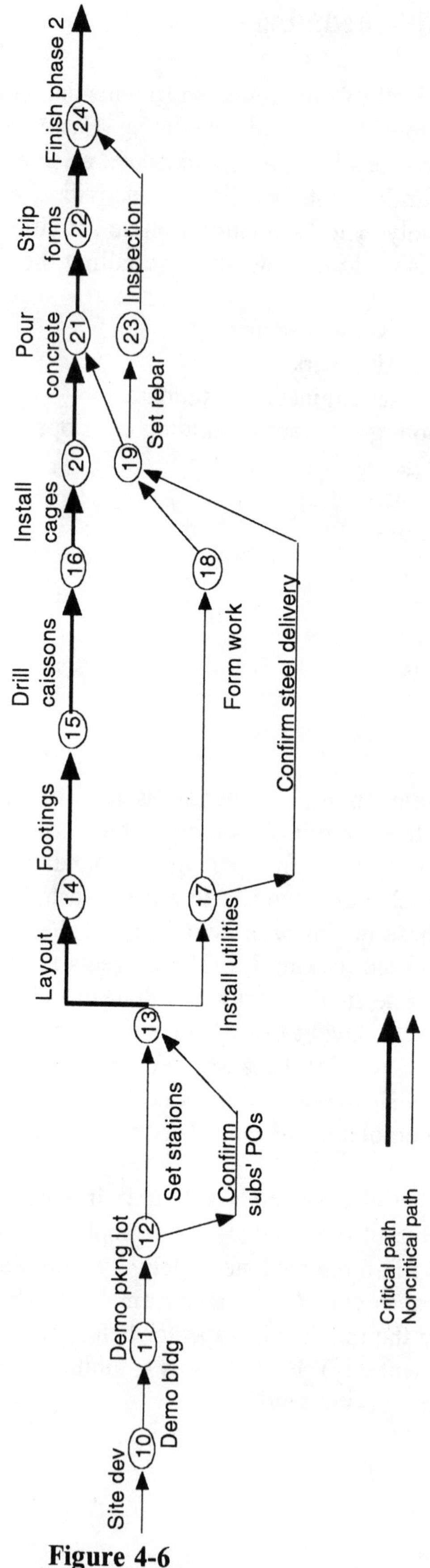

Figure 4-6
Logic Based Scheduling

The non-critical path for the foundation phase lies through activities 13, 17, 18, and 19. The critical path lies through activities 10, 11, 12, 13, 14, 15, 16, 20, 21, and 22. To decrease total phase duration in our network scheduling, we seek ways to shorten the critical path activities durations. If we could shorten the path through the Footings activity by more than seven days, the critical path would shift to the Form work path. If we could shorten the path through the Form work by three days, we would have two critical paths of (15) Footings and (18) Form Work running in parallel duration of ten days.

In network scheduling, when activities are combined in a network in which one activity shares a dependency on a processor activity with another activity, a parallel arrow is introduced into the network, as illustrated in Fig. 4-6, to show the interdependency relationship, as in the case of activities (17-19) and (19-24) to the critical path activities (13-21) and (21-24). A dummy arrow along either path, or subpaths, would indicate a constraint on that activity by a predecessor on that path. A dummy arrow representing a constraint differs from an activity arrow in that it does not represent time, only activity event dependency. All other arrows represent both time and dependency.

To illustrate the timescale difference between paths 1, 2, and 3, activity (17-19) can start up seven days later than activity (14-16) and still finish without extending the critical path of (15-21). Path 1 has five days of total float, path 2 has seven days of free float, and path 3 has ten days of free float. The free floats of (17-19), and (14-16) can be used to add to the total float of all the activities on the critical path.

By adjusting the activities start and finish events to match the logic-based schedule in Fig. 4-6, we begin to interlock the necessary components. In Fig. 4-7, the numbers in the circles represent the *i-j* numbers for the activities, based upon early start and late finish. The numbers under the lines represent the amount of time required in days to complete the activity.

By adding the final ingredient of timescale linkage, our network schedule now has controlling CPM. The numbers below or across from a circled i-j number represent days duration, with no float factoring. The time to reach starting event 17 from starting event 13 is seven days, and it is not on the critical path. This allows it to contribute its free float (if any) to the project's total float as needed for the critical path through events 14 to 24. By inspection of the timescaled network, we see that activity (17-19) has ten days of free float compared to activity (17-18), although both must be accomplished by starting event of activity 19. These types of parallel activities are also used as workarounds in the event of a bottleneck on the critical path.

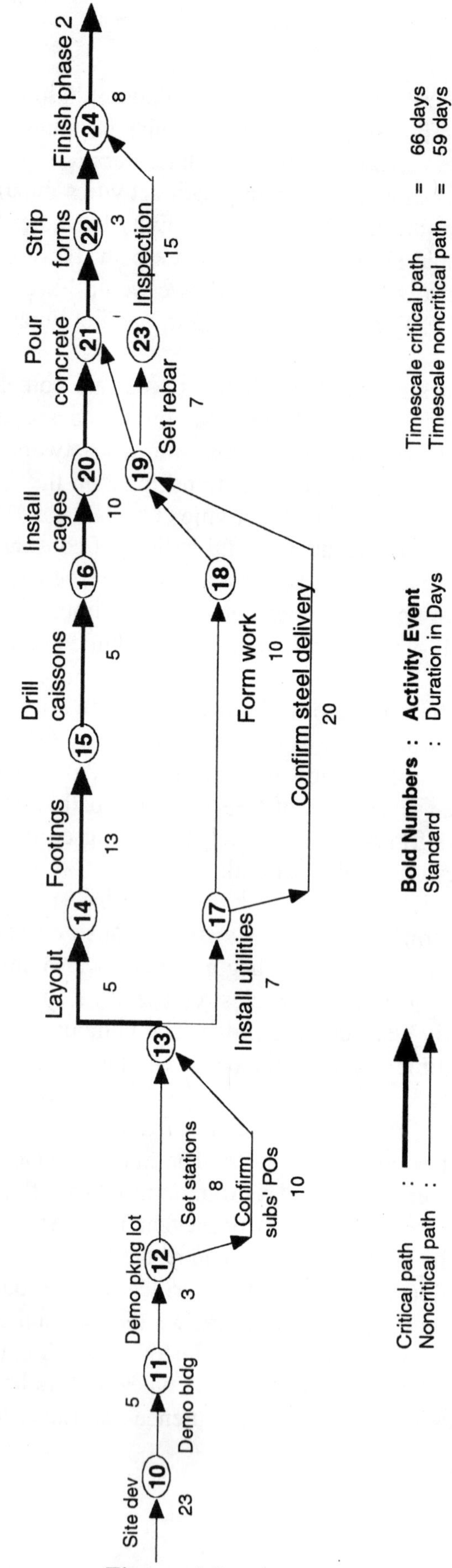

Figure 4-7
Logic Based Duration Scheduling

Examining the arrow diagram and the timescaled chart in Fig. 4-7 shows that the path through activities 17 through 21 cannot proceed past event 21 until the completion of activity 15. Event 21 must be delayed until the completion of event 19. Event 23 and event 21 are to be completed on the same day, so they are shown as happening in parallel. The path from event 16 to event 20 is a total of ten days, as they are in series, in contrast to the ten days total duration for both paths from activity (16-20) and from (18-19). Both paths converge again at event 24, which is constrained by activity 23.

Early and late start dates for each activity represent the earliest that activity can start. Those events combined with early and late finish dates for that activity (which represent the latest that activity can finish), are the durations which will change the critical path of the production. Events are further assigned a computer identification number, with the starting event number being the *i* number and the completion event being the *j* number. The *i-j* number then links the event to the network timeline as the relative cell address for the activity's data sort. When event numbers are assigned, the finishing event number at the head of the arrow must be greater that the starting event number at the tail of the arrow, such as event 19 leading to event 21. The computer reads the network schedule timeline as a mathematical sequence, and the *j* value of each activity must be greater than its *i* value for computations of program logic.

Without the use of float, any change in the start day of any event on the critical path will result in changing the completion date of the project. If activity (19) Set rebar is delayed, its delay would push back activity (21) Pour concrete, whereas activity (19) Set rebar and activity (20) Install cages are running in parallel. A typical CPM network involves hundreds of separate activities running in series or in parallel, all of which must have timeline flexibility in the network scheduling. Working back from event 24 on the network schedule in Fig. 4-7, it can be seen that activities (13-22), running in series, must be completed first, plus all activities (17-23), running in parallel, must precede activity 24.

Now we will take all the examples so far of CPM network scheduling and add the ingredient of time manipulation. As we said, without the use of float, any change in the start day of any event on the critical path will change the completion date of the project. Now, by adding the element of float and the flexibility of workarounds, we begin to gain control over the network timeline.

Chapter 5

Float

Types of Float

Float can best be described as scheduling leeway or slack time between production activities. In CPM, float is the number of days a non-critical path activity can be delayed before it becomes part of the critical path. All float types that we will look at in this chapter are in increments of working days. It is in the best interest of the prime contractor, subcontractors, the project management team, and the owner that the project have float periods. These float periods assure that every activity has some time flexibility. This window of cushioning can be opened or closed (accelerated or constrained) at a critical point to provide a cost-effective production timeframe. Float, properly used in CPM scheduling, is a major tool of time manipulation.

Float periods show that every project schedule, even fast-track, has some timeline flexibility. A non-critical path activity can be started a few days after it was scheduled to begin without delaying the critical completion date, whereas a critical path activity must be started immediately once the preceding critical path activity has been completed. Additionally, a non-critical path activity can finish later than the date it was scheduled for completion. However, if a critical path activity's completion is delayed, the total project will be delayed. In terms of constraint on a non-critical path activity, it cannot be delayed longer than its float period without moving to and becoming part of the critical path. By using float in CPM scheduling, time manipulation of non-critical path activities means accelerating or constraining their starting events a few days after they were originally scheduled, without delaying the total project's completion date.

When an activity has float time available, these extra days may be used to serve a variety of scheduling purposes. When float is available, the earliest starting event of an activity can be delayed, its finishing event extended, or a combination of both can occur. We do this through constraints and acceleration. Constraints are explained later in this chapter, acceleration speaks for itself. If we can string activities tightly together, obviously we'll save the most money. To successfully manage the schedules of noncritical activities and work items within the critical activities on the critical path, the project scheduler must understand the interdependent workings of float times and constraints.

Free Float

There are three basic types of float time, and other subfactoring types of float within those basic three. The major three are: free float, total float and negative float. Free float is the exact amount of days that any specified activity can be delayed without delaying the early start of the succeeding activity. As we learned earlier, events mark these days and therein the days of free float available on that activity. In schedule computations, free float is the least difference in days between the early finish event of an activity and the early start event of all following activities.

Total Float

Total float is the amount of time that an activity can be delayed without delaying the late finish event for the project's completion. Total float is shared with all activities. When an activity has a certain amount of total float, it can be used without tighter scheduling constraints affecting to all the other critical path activities.

Free float, however, is not shared with other activities as is total float, and accordingly free float provides the only true measure of how many days an activity can be delayed or extended without delaying any of the other activities. It must also be added as part of the total float of preceding activities. In Fig. 4-7 the network schedule arrow diagram is used to show a CPM network with *i-j* numbers shown in the circles and the numbers beneath or across from the arrows representing time duration required in days. Free float would be a portion of each activity's non-critical duration. Total float would be the sum of each activity's remaining free float that could be committed to the critical path. Total float is shown on the CPM summary report "Sort By Total Float/Late Start." It is computed twice by the computer; once plotted on the spreadsheet for early start and finish events, then again plotted for late start and finish events.

Line Float

Line float is amount of time slack available per line item on a computer spreadsheet. The line float time is easy to compute, because it is simply the difference between the early and late dates for an activity. It represents the available time between the earliest time in which an activity can be accomplished (based upon the status of the project to date) and the latest time by which it must finish for the project to finish by its event deadline.

Negative Float

When the critical path has been delayed and production is now behind schedule, the earliest starting event when an activity can begin is now past the latest time in which it can be completed to stay on schedule. This is known as negative float. As the activity has no float, its completion time is reduced to critical duration, and if it is behind, the difference between the early and late dates is less than zero. Negative float shows how far behind schedule the activity is, and if it is a critical path activity, it shows how late the project completion will be.

Distributed Float

As we will see later in the Business Combat chapter, float is sometimes used as a legal weapon by owners and contractors to approve or deny claims for liquidated damages in CPM contracts. Distributed float calculations are used in network scheduling to prevent float being used up by either party in a premature or unjustifiable manner, thereby deleting float from the total project. Distributed float allocates float to each individual activity by using PERT system duration averaging.

Low Float

The most critical work activities within the critical path(s) are assigned low float timing. Leeway is crucial to contractual completion date. Low float activities are listed first on the Total Float Sort, for primary attention.

High Float

Less critical work activities within the critical path(s) can be scheduled with more event timing leeway. High float appearing in later project activities is listed on the Total Float Sort for schedule recycling analysis.

Constraints

Constraints are delays placed upon an activity's starting event by practical limitations, preceding activity delay, or the project scheduler's option as a means of control over the network timeline by using types of float as time manipulation of that event. They are used as factoring tools in timescale computations of the network schedule. The only CPM fundamental in constraint on a non-critical path activity is that it cannot be delayed longer than its free float available. If a noncritical path activity is constrained beyond its float period, in CPM it therefore reaches negative float and automatically becomes part of the critical path. As the project proceeds, some of the activities that were previously on the critical path will no longer be there. Suppose there is a three day delay to framing. Originally, framing had one day of float. Now the CPM must be readjusted, because the total project has now been delayed two days. Framing has now become a critical path activity, and concrete has moved off the path.

Remember, constraints are those practical limitations that can influence the start of certain activities. The activity that involves the placing of reinforcing steel obviously cannot start until the steel is on the site. Therefore, the start of the activity of placing reinforcing steel is constrained by the time required to prepare and approve the necessary shop drawings, fabricate the steel and deliver it to the jobsite. In CPM, you treat constraints the same as activities and to represent them as event durations on the network diagram. These types of constraints prevent activities from starting until other preceding activities are sufficiently complete. Policy constraints are those methods the project team uses in the project's production to consume total float. Constraints are shown as dummy arrows on CPM network scheduling as dashed lines with zero elapsed time.

Float Example

The float of an activity represents the scheduling timing leeway available to be committed for total project critical path duration. When total float is available, the earliest start of an activity can be delayed, its duration extended, or a combination of both can occur as long as the late finish time is not exceeded. It is the window of timing that is available for that activity if the activities preceding it are started as early as possible and the ones following it as late as possible. When an activity has free float time available, this extra time may be utilized to accelerate or constrain succeeding activities. Critical Path Management uses float as a time manipulation tool, and the nomenclature and workings of float comprise essential knowledge of the project scheduler.

In summary, we have seen that the total float of an activity is the maximum time that its actual completion can go beyond its earliest finish time and not delay the entire project. If all the total float is used on one activity, a new critical path is created. The free float of an activity is the maximum time allotted by which its actual completion date can exceed its latest finish event without affecting either the overall project completion or the times of any subsequent activities. If a work item within an activity is delayed enough to consume an activity's free float, if it still completes on time, the activities following it are not affected and they can still start at their earliest start times.

If we took our example of the CPM network schedule in Fig. 4-7, and doubled the phase in size to two quarters from January 13th to July 14th as in Fig. 5-1, the network schedule easily shows how days of scheduling leeway between activities could be used along the critical path to shorten activities duration if needed. Float is computed by certain factors determined by the network time, as in Fig. 5-1, as well as the unit factors of the Sort By Total Float/Late Start.

To illustrate those factoring units of float, we need to identify them each as in Fig. 5-2, our Sort By Total Float/Late Start summary report. The sort shows data columns for each factor of float, which are:

- The activity's description
- The activity's i-j number for computer identification
- The activity's estimated duration versus remaining duration
- The percentage of activity completion
- The activity's actual start and finish
- The activity's early and late start
- The activity's early and late finish
- The activity's remaining free float
- The activity's available total float

The computer determines each activity's float by computations between i-j numbers of events and corresponding timelines. In addition, we can determine the critical path in this phase runs from event 10 to event 24, the completion time for critical activities, early start and finish dates for each activity, free float, and total float available.

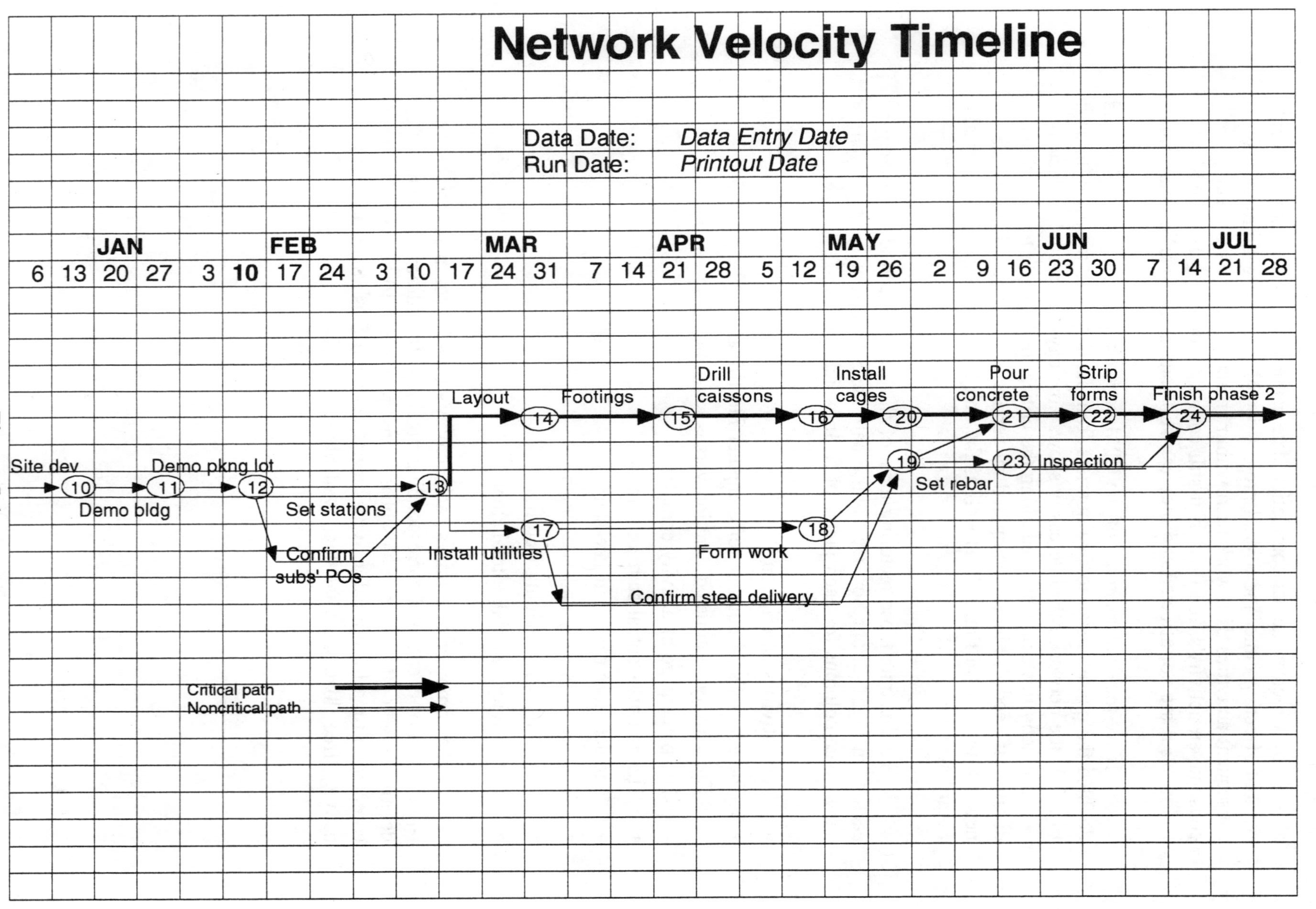

Figure 5-1
Durations Timeline

If we use the job logic of two sequential activities, such as placing reinforcement steel for a concrete slab and the succeeding activity of pouring the concrete, we can examine them using the factors in Fig. 5-2 to calculate float. If these activities are not on the critical path and the difference between the early and the late start dates is a total of five days, how much total float does the rebar activity have? (Assume no other activities have constraints on these activities.)

The answer is five days: the difference between the early and late start, or the early and late finish. Now, how much total float does the pour activity have? The answer is five days also, calculated from that activity's duration. Then together, how much total float does the entire sequence of activities have? Ten days? No, because total float cannot be added; the sequence of activities still has only five days. What if the rebar installer is two days late in starting? As of the time started, the activity subcontractor has only three days of total float. The prime contractor has also been reduced to three days of total float. If the placement of rebar finishes three days late, the prime contractor has no float left in the schedule and the project is on critical path.

In float computations within CPM scheduling, each activity does not have total float individually, but all the activities in all of the phases share the same total float time made from their individual free float. Phases of activities have total float, and if it is consumed by one activity, float no longer exists in that phase for any activities. Not a good situation for any project scheduler, and one that the pros avoid like the plague.

FastPro CPM Scheduler

Your Company Name Here				**Sort By Total Float/Late Start**				
Prepared For:	***Your Client's Name Here***							
Project:	***Project's Name Here***					**Page: 1 of 3**		
File Name:	*Your Computer File Name Here*			Data Date:	*Data Entry Date*			
Description:	*Project Description Here*			Run Date:	*Printout Date*			
Our Invoice:	***Your Company's Job Number***							
Client Invoice:	***Owner's Job Invoice Number***							
Phase 1								

I Node	J Node	Activity Description	Duration Est	% Comp	Actual Start	Actual Finish	Free Float	Total Float
		Subsurface Investigation						
		Instructions To Bidders						
		Information Available To Bidders						
		Bid Forms						
		Supplements To Bid Forms						
		Agreement Forms						
		Bonds And Certificates						
		General Conditions						
		Supplementary Conditions						
		Drawings & Schedules						
		Addenda & Modifications						
Phase 2:		**Specifications**						
		Subsurface Investigation						
		Allowances						
		Measurement & Payment						
		Alternates/Alternatives						
		Coordination						
		Field Engineering						
		Regulatory Requirements						
		Abbreviations & Symbols						
		Identification Systems						
		Recordation Systems						
		Reference Standards						
		Special Project Procedures						
		Project Meetings						
		Submittals						
		Quality Control						
		Construction Facilities & Temp Controls						
		Material & Equipment						
		Starting Of Systems/Commissioning						
		Contract Closeout						
		Maintenance						
Phase 3:		**Commencement: Site Work**						
		Subsurface Investigation						
		Demolition						
		Site Preparation						
		Dewatering						
		Shoring & Underpinning						
		Excavation Support Systems						
		Cofferdams						
		Earthwork						
		Tunneling						
		Piles & Caissons						
		Railroad Work						
		Marine Work						
		Paving & Surfacing						
		Piped Utility Materials						
		Water Distribution						
		Fuel Distribution						
		Sewer & Drainage						
		Restoration Of Underground Pipelines						
		Ponds & Reservoirs						

Figure 5-2
Sort by Total Float/Late Start

Chapter

6

Network Velocity Diagramming

Phase Scheduling

Phase scheduling refers to combining all the divisions of activities (designated as phases) into a multiple critical path velocity diagram. This type of network scheduling is complex, but can be made much more accurate by using computer systems. On commercial projects with 10,000 or more activities, CPM computer programs begin using ten multiple critical paths. Multiple critical path project scheduling software uses dedicated program logic in phase scheduling of multiple critical path management. In phase scheduling, precedence diagrams use float to act as an activity linkage cushion in the network schedule. Normally, the prime contractor divides the total project into different activities in a schedule parallel to the owner's. An industrial project typically will have thousands of activities, with each subcontractor performing a different activity, each of these having subactivities and tasks within them. Then the owner's project scheduler determines the critical activities that must be completed before the other activities can be started.

These types of job logic constraints are a function of phase scheduling CPM management. For example, usually site work must be completed before concrete work can begin, these activities running in series. Conversely, plumbing and electrical activities can be performed simultaneously, these activities running in parallel, as neither is typically dependent on the other. These activities can be in the same phase scheduling, because the subcontractors can work on them at the same time. By grouping activities with similar job logic we develop macroactivities that can be scheduled in phases. By stepping up to management at the macroactivity level, the sorts make large units of data manageable.

Remember, job logic constraints always dictate the path(s) of the CPM network schedule. Because excavation has no prior constraints, it can be performed first. Once it is completed, the formwork, plumbing, and electrical, which were constrained by the need to complete excavation first, can now start. The critical path, the longest path on a velocity network, consists of those activities that will cause a delay to the total project if they are delayed. In most projects the primary activities, such as excavation, formwork, concrete pour, and roofwork, are on the critical path. A delay to any of these activities will delay the entire project. In contrast, plumbing and electrical activities are usually not on the critical path. Their delay, up to a point, will not delay phase scheduling within the total project.

Multiple Critical Paths

A network with phase scheduling of ten critical paths of equal duration and event length would have varying levels of activities running in series, and others running in parallel. In such a network, the project team would first develop a primary critical path through the ten paths. Then they would adjust the durations of the other activities to create secondary paths that, together, would make the ten-critical-path condition. Normally, about twenty percent of project activities will continually fall on the critical path, so CPM is a management-by-exception technique. When more than twenty percent of the activities fall on the critical path, multiple critical paths need to be established or the schedule will no longer be achievable in its present form. Very little float is available in a multiple critical path network, and the owner is subject to claims any time it delays the prime contractor on any primary or secondary critical paths.

The contractor, in this situation, is in an adversarial position through clauses in the schedule specifications that direct the contractor to redraw the network any time it is behind schedule on the primary critical path. So it is in the contractor's best interest to eliminate float and back-charge the owner through claims for delay. Often this will last for the duration of the project.

This tactic in multiple critical path schedules, of eliminating as much float as possible from the network, puts the owner at a disadvantage by not having cushions for delays. The project moves faster but at a higher cost, due to claims made by the prime contractor at project closeout. Now that you see both perspectives, you can understand how, as a project scheduler, your use of different types of float in your phase scheduling can either benefit the owner (if that is your client), or the contractor (if that is your client). Another profit area for the hot scheduler.

In network velocity diagram sorts, as with all summary report printouts factored by the computer on a sorting by *i-j* node numbers, the order in which the activities appear on the list has no relationship to the job logic sequence. This is the computer linkage between the activity on the diagram and the computer data on that activity. Velocity indicates the amount of forward momentum in the timeline, and time manipulation is the tool we use to either accelerate or retard that forward progress in collective or individual activities. This tool is applied through the proper use of free float and total float. Adding these controls to phase scheduling in our network diagram gives us the foundation of network velocity diagramming. Adding event velocity to network velocity diagramming gives us the foundation of fast-tracking, which is today's production scheduler's "jump start" option, as we'll see in Chapter 7.

Bar Charts

The bar chart was invented in the late fifties by Henry Gantt. He was working as an engineer on a capital works job and developed the bar chart as a graphic representation of the project schedule he proposed. It was so successful that soon everybody was using them for tracking vertical components through a horizonal timeline, which was applicable to every business in existence. Bar charts are now used in all industries for easy and quick visual representation of data. As construction scheduling became more sophisticated, the bar chart was used in combination with hand spreadsheets, to schedule and track construction projects.

Computerized CPM scheduling also produces a bar chart printout sort, which is sorted from the activities' early starts. Bar charts are included in the weekly meetings, as progress reports of scheduled progress versus actual progress. These bar chart sorts are an essential tool for presentation of network scheduling details in the CPM project schedule to the owner and the project team.

Bar charts are effective at activity scheduling and tracking as a job logic diagram. They are cost-effective scheduling tools for smaller projects and on residential work. In commercial and industrial work, schedulers and site supervisors normally have bar charts covering one wall of the job trailer. Lots of different colors of highlighter going everywhere on the diagram. No one but a scheduler has the time or the desire to compute the variables.

Bar charts, however, do not show all the interdependencies between activities, whereas CPM schedules show the dependence (constraint) of one starting activity and the finish of the activity preceding it. Further, bar charts to not allow for variable float control at those activities' events. There are also limits to the number of activities, usually around fifty, that can be tracked on a bar chart before the chart becomes overloaded and the milestones within the bar chart schedule miss their marks.

S-Charts

S-curve charts, known in the trades as S-charts, add timeline momentum to bar charts and can be thought of as flow charts. These are the types of graphs that use coordinates to plot flowlines, as in the example in Fig. 6-1. These are scaled in a tighter timeline than the bar chart, to provide more scheduling control over subtasks within activities. Bar charts are typically structured with a one-month timeline, whereas S-charts are structured with a one week timeline, which opens up the activity and its subtasks to greater detail and control. The project scheduler now obtains the margins between work items necessary to establish workable free float.

S-charts show the interrelationships of sequential and parallel activities more clearly than do bar charts. The manipulation of the starting and finishing events is critical to effective CPM. Even work items within subtasks within activities should be broken out, with individual computer numbers for dedicated cell address data storage. All activities are made up of smaller activities than need attention and control. For example, if we have an activity "Obtain building permits," it is entered and computed as one activity, whereas anyone experienced in pulling permits will tell you that obtaining building permits is far from a one-step operation. Many subactivities are involved in the permit process alone. Each of the activities have subactivities involved, which are not shown here for clarity of concept.

Once the scheduler has identified these work items at the microcosom level, the activities listed on the primary report file (Sort by Activities) are then broken out from the Sort by Activities timeline and transposed onto the accelerated S-chart timeline. This, in effect, shifts the production network into double-time. Professional ball players have reported that, when they are really in the groove and doing well, things seem to slow down a notch and they can really control their timing. I once heard Matt Williams remark that he started his home run streak when "time seemed to go into slow motion and I could see every stitch lace on the baseball quite clearly as it was coming at me." So he focused better and connected. This is exactly what happens at this transposition point of superimposing the Network Timeline S-chart on the Sort by Activities bar chart. Here is where *you* focus and connect. This is the point in your development of the schedule where the double-timing opens up details on subtasks and work items within the activities, and thus, windows of float. Superimposed events and milestones begin to interweave, and the job logic of the network schedule becomes workable.

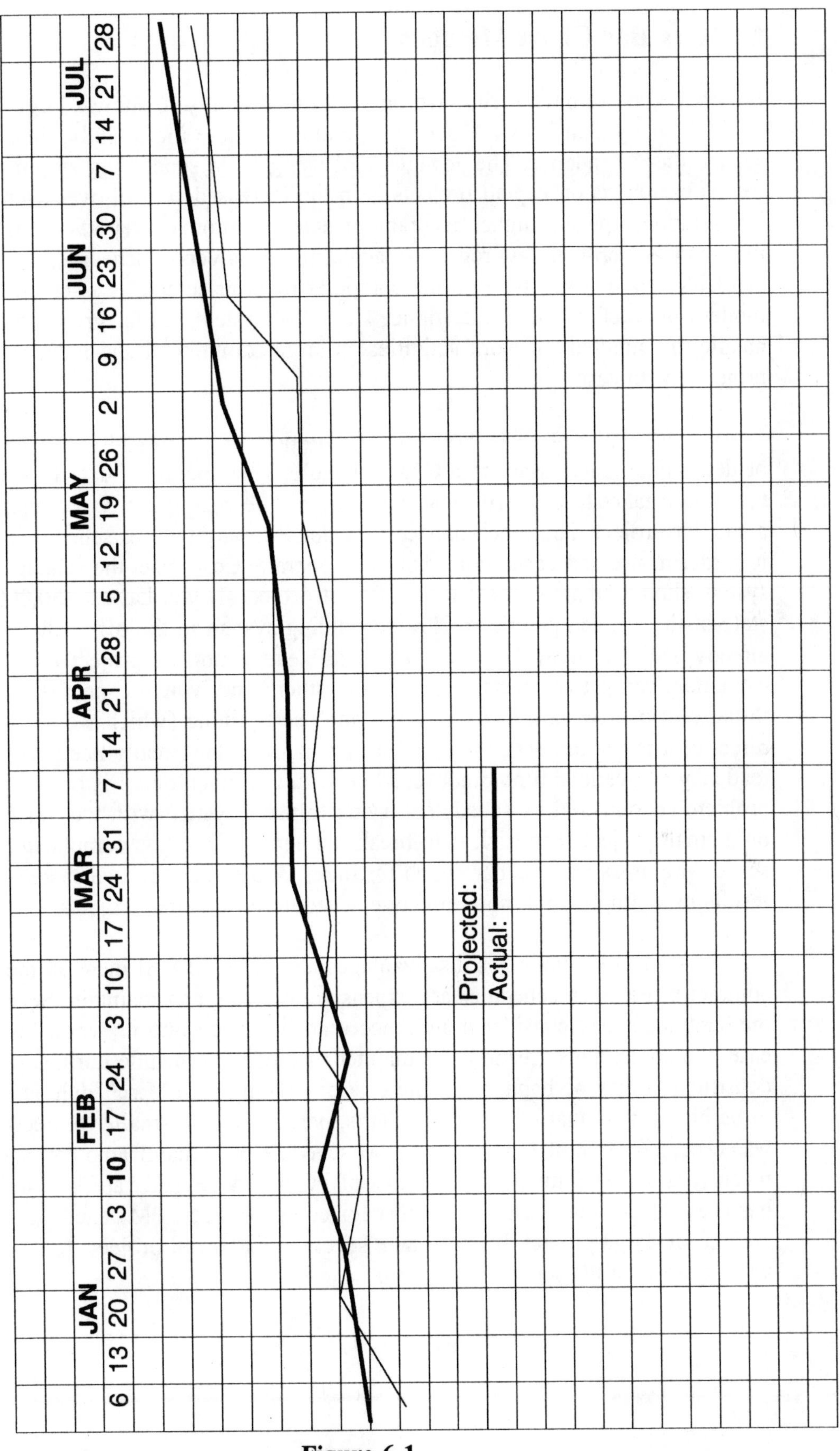

Figure 6-1
Flowline by Plotted Coordinates

CPM vs Bar Chart Methods

Bar charts, as we have seen, have only a limited ability to show a large project's many detailed work activities, and even more importantly, their time-scaled interdependence. The advantage of using CPM scheduling systems versus bar charts on capital projects is in the increased control over event interrelationships and higher accuracy of detailed summary reports (sorts). The instantaneous critical data regarding actual progress versus schedule, readily available at any given time on the computer monitor screen, is essential for any modern production under a deadline. The CPM system has far greater ability to handle the hundreds of work activities on large commercial and industrial projects, with ease.

On smaller projects, however, it is important not to add more activities or detail than is necessary to a CPM schedule. The amount of data flow and the time it takes to deal with that data may not be cost-effective to generate on a smaller project. Too much unnecessary detail in any type of schedule makes it harder to use and costs more money to operate. Cost-wise, the scheduling system should be appropriate to the size of project. If you have a $50,000, state-of-the-art, computerized CPM scheduling system in the office that's already paid for, great! Go out and schedule every monster project your estimators can get contracts on. If, on the other hand, you're concerned with the initial cost of new hardware, software, and user training (which would be a direct cost to the project), then don't buy what you don't really need. This tendency to overload a nominal schedule with too much detail is a common problem encountered by schedulers trying to use a system with too much power on a small project. The trick is to break out some of the less complicated scheduling areas and use bar charts for them. A blend of the two systems often results in a simpler and more effective schedule for smaller projects.

A CPM schedule has disadvantages also. First, it will increase the total contract price. Such schedules are expensive to create and maintain. Second, the contractor may consider it an unnecessary intrusion into its work. In such a case, the contractor's creation and maintenance of the schedule during construction may be haphazard. The cost of running a CPM schedule is likely to be higher than that of using bar charts, particularly on smaller projects. That was especially true of running the CPM schedule on a mainframe computer. In recent years, the relatively modest cost of PC hardware, software, and training has enabled us to expense-off the cost for computerized CPM scheduling on a medium-to-large project. It might take several small-sized projects to cover the writing off of a PC scheduling system.

Arrow Diagramming

Once the scheduler has identified the job logic, the schedule is plotted in a flow chart by using arrow diagramming. In arrow diagramming, the basic unit is a work activity that occurs between two events. The events in arrow diagramming are known as nodes. The events or nodes are numbered sequentially, and the activity is identified by the beginning and ending event numbers. Those numbers are also designated as the *i-j* number, as shown on the computer summary report Sort by I-J Numbers.

Each activity has a projected duration necessary to complete the activity. Each activity's estimated duration is determined by the parameters of the work items involved. The best source of data for factoring here is your company's historical data, available from previous similar projects. If there are no historical data on the activity, the project scheduler uses cost estimate publications, such as Lee Saylor or Means Cost Index, and uses area adjustments for geographical cost factoring. Experienced project schedulers will also seek a best estimate of the activity's duration from input from the subcontractors most experienced in performing that activity. The function of the arrow diagram is to determine the longest time path through the diagram, which is the critical path.

In Fig. 6-2, the critical path for project completion passes through activities 1-2, 2-3, 3-4, 4-6, 6-7, 7-10, 10-11, 11-12, 12-13, 13-15, 15-18, 18-19, and 19-23, for a total of 190 working days. Path two (4-21, 21-22, 22-23) and path three (4-16, 16-17, 17-21) are the same duration of 120 working days. The fourth path (4-9, 9-11) lasts only 35 days, and then rejoins the critical path, having float of five days. Path five (4-8, 8-20, 22-23) lasts for half the duration of path three and shows float of ten days. Keep in mind, however, that this arrow diagramming shows constraints but not accelerators.

To accelerate activities' durations we manipulate the early and late start events and the early and late finish events. These have yet to be added to the CPM arrow diagram in Fig. 6-2.

Precedence Diagramming

The next level of network diagramming is called the precedence diagraming method, or PDM. In this method, the activities are represented on the network as boxes instead of arrows. We call these boxes *nodes*, and in PDM the arrows between the nodes only indicate the established job logic interdependencies among the activities. Nodes may be represented on the network diagram either as boxes or circles.

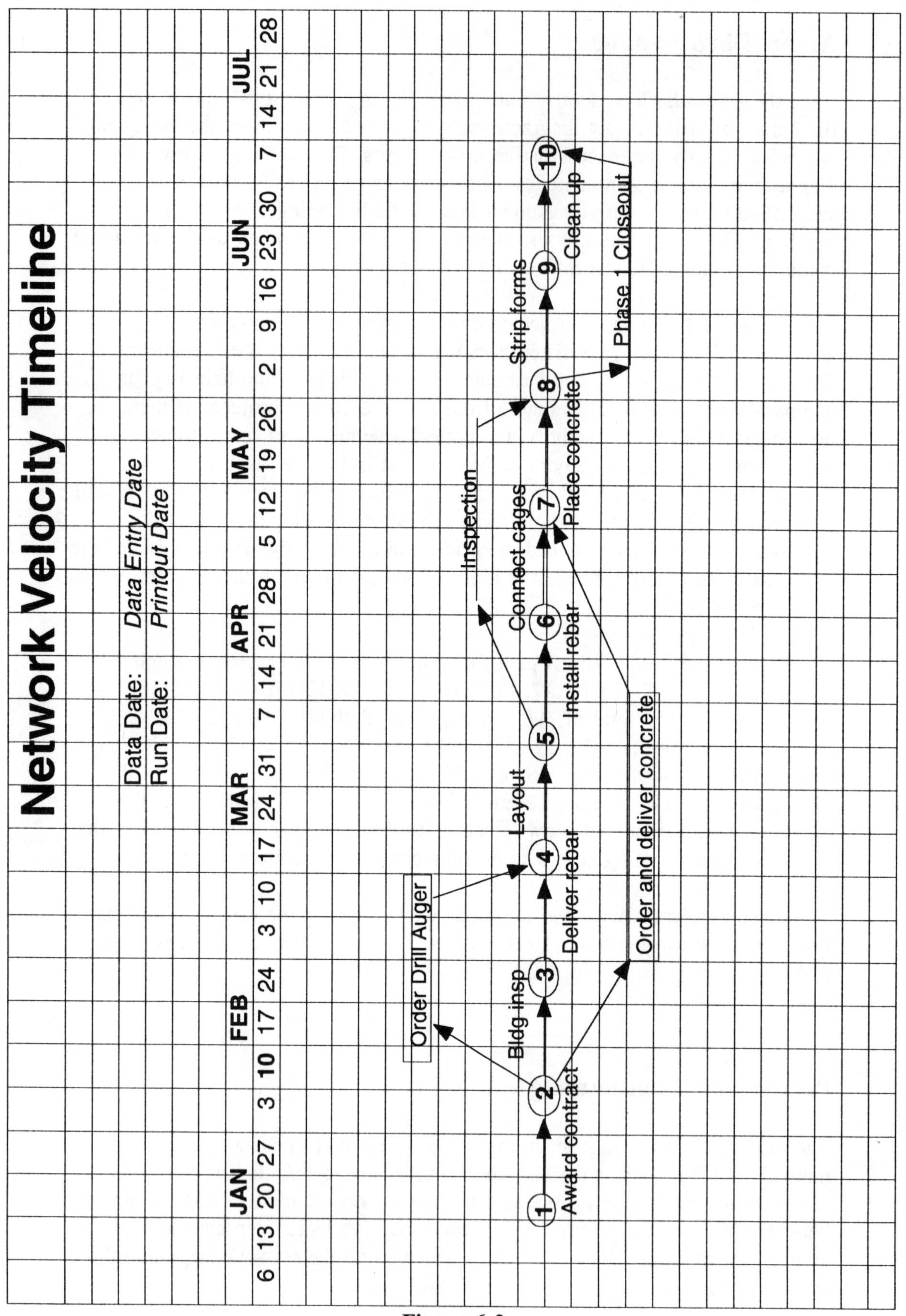

Figure 6-2
Arrow Diagramming

PDM diagramming shows the activity number in a block and not on the arrow line, as shown in the arrow diagram in Fig. 6-2. The larger rectangles are better suited for detailed activity descriptions and duration events than are circles with numbers in the vicinity, as in an arrow diagram. Instead of a single *i-j* number appearing in the circle, in PDM the activity number also appears in the block as further identification and data detail. The nomenclature of an arrow diagram and a node diagram are the same, as are calculating float times and the early and late start dates for all activities in the logic diagram.

A PDM diagram is easier to read than an arrow diagram, because the connecting line is shorten between nodes. The main advantage of PDM is its ability to show the constraints an activity has on its succeeding activity. CPM programs that run on Activity-on-node formats are combinations of both ADM and PDM systems. Scheduling personnel typically like to have both ADM and PDM diagrams displayed, so they can visually check the logic of the diagram in detail as the project proceeds. The project managers need to refer to it to resolve production bottlenecks. Usually, however, project managers work from the early and late start dates, increasing float sort and milestone sorts. Site supervisors typically use the bar chart sorts for day-to-day supervision of the construction work.

In PDM diagramming, the length and direction of the arrows show only the flowline of interrelationship. Often arrows are deleted altogether, and only lines connect the nodes. The line between two nodes indicates the relationship between them; these lines are connectors only and do not represent timescale, because they are all of zero time duration. PDM diagramming thus eliminates the need for dummy arrows. Different notations can be made in the node, as the project scheduler needs to show a wider range of events' interdependencies or more detailed data. This gives more flexibility to event scheduling when building the job logic diagram.

As was mentioned previously, typically about twenty percent of project activities will continually fall on the critical path, so CPM is a management-by-exception technique. When more than twenty percent of the activities on a PDM or Activity-on-node diagram fall on the critical path, it is a sign that the network schedule is falling behind in its timing and needs to be recycled immediately. Recycling the schedule is done at the end of the month, but needs to be done immediately if your short-term milestones are not being met.

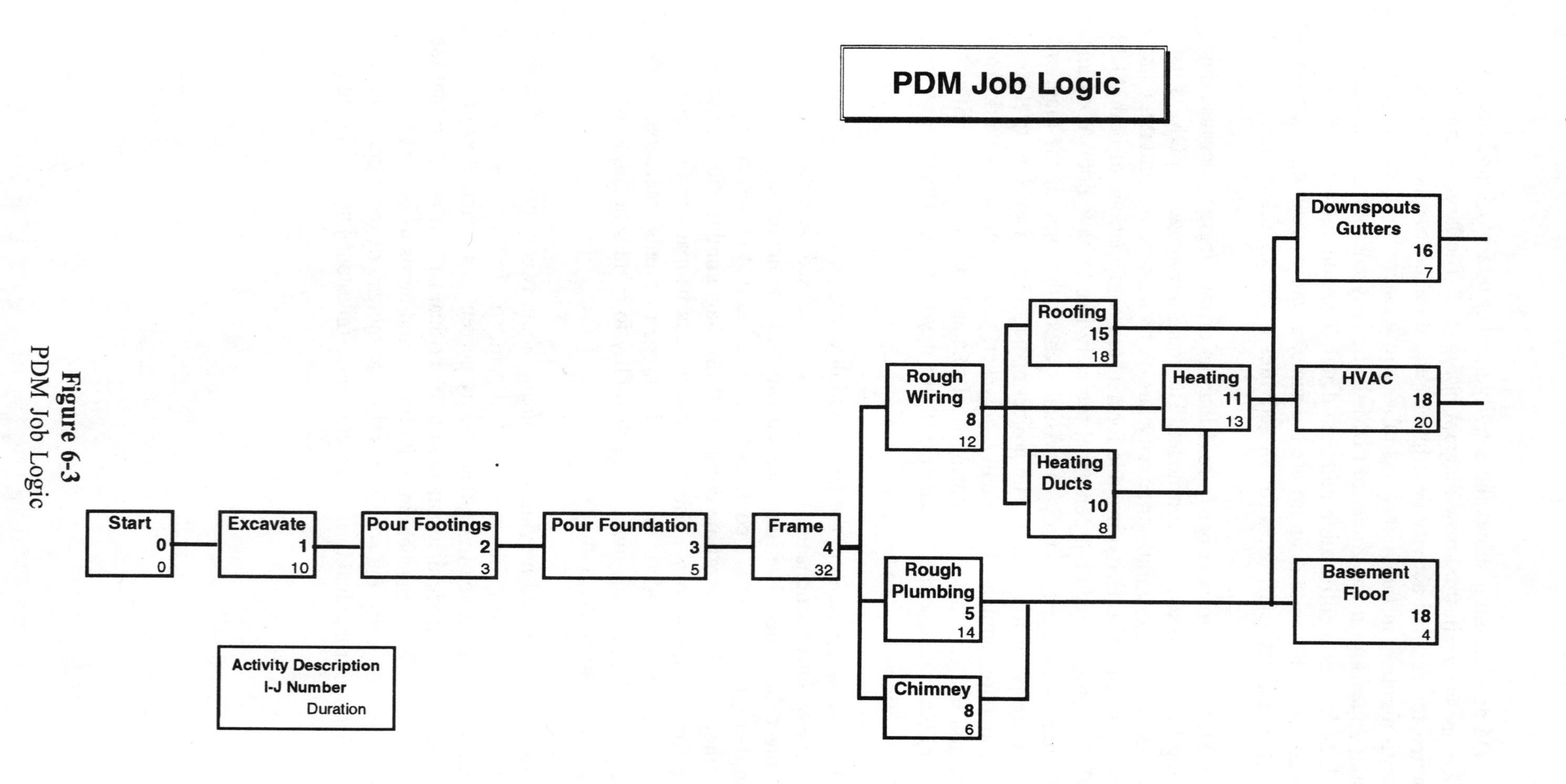

Figure 6-3
PDM Job Logic

Figure 6-3 shows typical PDM job logic on a residential project. Although the two methods of producing ADM and PDM diagrams produce different appearing networks, the production job logic remains the same. In a precedence diagram, the activity that would be represented as an arrow in ADM diagramming is now shown by the rectangular activity box (node).

In ADM, a dummy arrow differs from an activity arrow in that it represents a constraint, while all other arrows represent both time and dependency. When four or more activities converge in one intersection event, a dummy arrow is introduced into the diagram network to indicate where an activity has a starting event constraint dependency on a preceding activity. In PDM diagramming, however, the four or more activities would be best-time-linked in a series according to established job logic.

ADM vs PDM

ADM scheduling requires knowing the timing significance and proper usage of dummy arrows, and mistakes using constraints or accelerators in the wrong place are disastrous in a CPM network schedule. Therefore, the standard in the industry is use of a PDM network by the site supervisor and field personnel rather than arrow diagramming. The nodes' ability to expand to include any pertinent data also makes it easier to modify the network diagram with PDM.

PDM is popular in the business for these reasons, and is in wide use through the modern construction industry, and especially in it's CPM version activity-on-node format, which we'll look at next. However, ADM also has a distinct advantage over PDM. The major difference between the two, and where ADM has the time-manipulation edge, is that there are no events in PDM box node diagrams, which makes it impossible to use early and late start and finish events to interrelate events between activities networks.

Activity-on-Node

Activity-on-node diagramming is a combination of both ADM and PDM that produces a superior precedence diagram format for scheduling that is superior to either. This format is a variation of PDM diagramming that uses dependency arrows with no constraints (dummy arrows). In activity-on-node diagramming, the activities are shown as horizonal bars, their lengths relative to their duration. Activity nodes run from event to event and a single, narrower line following each activity node represents float. When run on the computer network timeline, the resulting diagram becomes a time-scaled network.

Activity-on-node uses a format wherein the horizonal activities bars contain the activities' respective i-j numbers, which the computer uses to best-time sequence the *j* node (finishing event) of the preceding activity with the *i* node (starting event) of the succeeding activity, and thus establish the dependency lines that determine job logic. Sometimes, if constraint must be shown, a dummy activity is used to show the activity event dependency. Conventional CPM network scheduling programs, by far, use activity-on-node formats for network timeline computations. Some programs offer what they term "dependency bar chart scheduling," which is a similar activity-on-node format.

Fig. 6-4 is an example of activity-on-node diagramming, showing activities interrelationships, dependencies, job logic, activity durations and available float, with constraints noted as dummy arrows. Note the similarities to PDM, with the added benefit of trackable float.

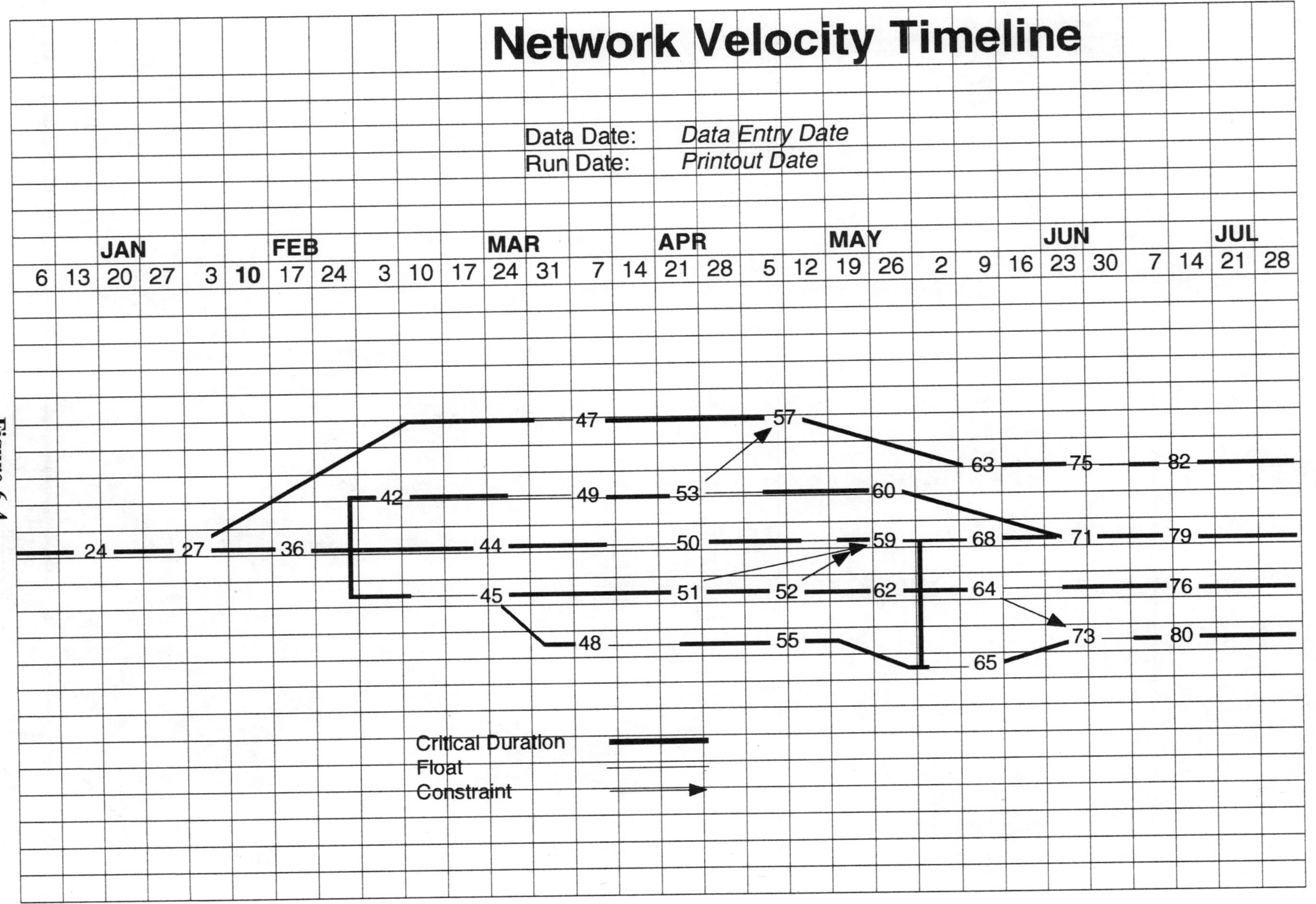

Figure 6-4
Activity-on-Node Diagramming

Chapter

7

Fast-Tracking

Phased Construction

Phased construction, or fast-tracking, is a modern variation of the previously examined CPM techniques. In fast-track projects, prearranged phases of the project are started prior to main project commencement. The big advantage to the owner in fast-tracking is that production can begin before the total project design is completed, thereby incurring less interest cost on the construction loan by completing the project sooner. An owner who is not yet due for the first draw payment from the bank does not yet have use of bank construction loan funds, and will have to draw on personal credit lines to cover initial costs of commencing the project. This additional debt load taps many owners thin for project commencement cost contingencies, and further reduces the owner's credit line to cover those contingencies. Through fast-tracking, however, construction can begin with certain activities completed and ready for initial draw progress payment from the bank's construction loan funds. The project is completed sooner, the owner has income from the capital project sooner, and the construction loan is paid off sooner at a lower cost. Same project, but more income and less cost. This type of project execution equates to sizable profit to the owner, but it must be remembered that the economic viability of fast-track project execution depends on the amount of the design completion available when the project schedule is being developed.

Fast-tracking has proven itself to be a viable project execution option in commercial construction and is being touted as the next wave of the future. I have my doubts about that point, but fast-tracking does work extremely well in certain types of similar multilevel types of construction projects. The primary advantage in the fast-track method is that construction can begin while design is still being worked out. This should mean that the project will be completed sooner. Obviously owners like fast-tracking. It makes them money and cuts their debts sooner. It will no doubt continue to be a real contender in the field of future professional project scheduling executions. Phased construction has become so popular in current project scheduling that CMI and similar corporations are now offering full-program courses in fast-track CPM. These types of nontraditional project production execution methods are indications of the changes yet to come for the professional CPM project scheduler.

Fast-track project scheduling has evolved as an efficient scheduling system for obtaining equal or higher-quality construction at the lowest possible price, and in the quickest turnaround timeframe. But, as with all things, there is a downside to fast-tracking that owners are frequently unaware of. A study by the U.S. General Accounting Office of the use of CPM Construction Management by federal agencies noted cost overruns with fast-track project execution in some projects, due to contractor delay claims. The study further found that if fast-tracking is used in a single contracting system, phased construction leads to cost-plus contracts by prime contractors, which means that the contract price will be tied to costs incurred no matter what numbers the design finishes with, since there are too many unknown variables to lock in a final price when the contracts are signed.

It also noted the difficulty of using CPM concepts with untrained users in federal construction, and reports that those projects experienced time delays, some exceeding six months. Conversely, however, the study also found that those projects using CPM fast-tracking with personnel trained in CPM techniques had cost decreases that averaged fifteen percent. Obviously, then, CPM fast-tracking will achieve its objective advantages only when applied by those who know how to operate the system. Intuitively one can grasp here that the reciprocal then must also be true: a CPM fast-tracking system is not cost-effective when used by personnel untrained in its use.

Design/Build

Often a fast-track project schedule is used in combination with a design/build system. This means that the designer and builder are the same entity. It also means that the prospective bidders cannot be given a detailed set of plans and specifications on which to make their bids, as design is not yet complete. So contractors bid by square footage costs added to a profit margin percentage, and contract by cost-plus contracts tied to whatever actual costs are incurred, regardless of projected costs. The advantage to the design/build system, from a project scheduling standpoint, is its speed of project completion, because the design/build system uses phased construction already in progress as its scheduling basis. By having materials or inventory in advance, contractors can start the project with modular sections of their activities already completed.

Adding pre-commencement activity completions and events timing controls to our network diagram gives us the factors of modern fast-tracking CPM. Design/build systems use events time manipulation to either accelerate or retard forward progress in collective or individual activities, producing forward momentum in production scheduling. This tool is applied through the proper use of free float and total float, in parallel activities as well as series activities. Adding these events' time-manipulation controls to our network velocity diagram gives us the working foundation of CPM fast-tracking. This is a modern scheduling technique of today's professional project scheduler.

Let's look at the basic elements of a fast-track system. Fig. 7-1 represents the first element, showing a bar chart schedule of a typical home-building project, scheduling the phases of that project. I have made this example bare-bones and overly simplistic for the purpose of illustration. The project's phases are:

1. Bid out design & plans.
2. Approve design & plans contract.
3. Obtain building permits.
4. Bid out construction.
5. Negotiate contractors' bids.
6. Complete any required geotechnical site surveys and soils testing.
7. Negotiate project schedule with best bidder.
8. Approve and award winning general contractor's bid.
9. Obtain contractors bonds, Workers Comp & liability insurance.
10. Begin site work.
11. Final inspection site work.
12. First progress payment.
13. Begin foundation work.
14. Final inspection foundation work.
15. Second progress payment
16. Begin framing work.
17. Pre-final inspection framing work.
18. Third progress payment
19. Begin electrical, plumbing and HVAC rough out work.
20. Final inspection framing work.
21. Fourth progress payment.
22. Pre-final inspection electrical, plumbing and HVAC rough out work.
23. Fifth progress payment.
24. Begin interior work.
25. Pre-final inspection interior work.
26. Sixth progress payment.
27. Final inspection electrical, plumbing and HVAC work.
28. Seventh progress payment.
29. Final inspection exterior and roof.
30. Eighth progress payment.
31. Finish out punch list.
32. Finish out changes & claims.
33. Final total building project inspection.
34. Project close out.
35. Contract close out.
36. Final payment & release of retention.
37. Obtain release of liens and file Notice of Completion.

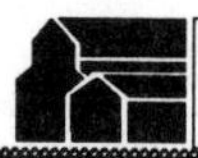

FastPro CPM Scheduler

CERTIFIED CONSULTANTS

Sort By Activities

Prepared For: **Power Engineering Corporation**
Project: **Stanford University**
File Name: cc/pe/stndford/activties
Description: Cancer Research Building
Our Invoice: **95-097-PEC**
Client Invoice: **CRB-102**

Data Date: 22 Dec 93
Run Date: 14 Jan 94

Activity Number	Activity Description	Original Duration	Remaining Duration	% Comp	Early Start	Late Start	Early Finish	Late Finish
General Conditions								
00010	Subsurface Investigation	30	0	100	20 Nov 92	24 Nov 92	10 Dec 92	12 Dec 92
00100	Instructions To Bidders	12	0	100	13 Nov 92	18 Nov 92	26 Nov 92	28 Nov 92
00200	Information Available To Bidders	2	0	100	28 Nov 92	02 Dec 92	04 Dec 92	08 Dec 92
00300	Bid Forms	1	0	100	10 Dec 92	12 Dec 92	11 Dec 92	15 Dec 92
00400	Supplements To Bid Forms	3	0	100	18 Dec 92	20 Dec 92	20 Dec 92	24 Dec 92
00500	Agreement Forms	0.5	0	100	10 Dec 92	12 Dec 92	11 Dec 92	15 Dec 92
00600	Bonds And Certificates	2	0	100	12 Dec 92	14 Dec 92	14 Dec 92	18 Dec 92
00700	General Conditions	0.5	0	100	17 Dec 92	19 Dec 92	17 Dec 92	20 Dec 92
00800	Supplementary Conditions	0.5	0	100	17 Dec 92	19 Dec 92	17 Dec 92	20 Dec 92
00850	Drawings & Schedules	1	0	100	18 Dec 92	19 Dec 92	17 Dec 92	20 Dec 92
00900	Addenda & Modifications	1	0	100	19 Dec 92	21 Dec 92	20 Dec 92	25 Dec 92
Subtotal:		**53.5**	**0**	**100%**	**29**	**30**	**10**	**17**
Phase 1:	**Specifications**							
01010	Soils: Reports & Remediations	10	0	100	20 Nov 92	24 Dec 92	10 Jan 93	15 Jan 93
01020	Allowances	1	0	100	10 Dec 92	12 Dec 92	11 Dec 92	15 Dec 92
01025	Measurement & Payment	1	0	100	17 Dec 92	19 Dec 92	17 Dec 92	20 Dec 92
01030	Alternates/Alternatives	0	0	100	17 Dec 92	19 Dec 92	17 Dec 92	20 Dec 92
01040	Coordination	0.5	0	100	17 Dec 92	19 Dec 92	18 Dec 92	20 Dec 92
01050	Field Engineering	3	0	100	19 Dec 92	20 Dec 92	20 Dec 92	24 Dec 92
01060	Regulatory Requirements	1	0	100	26 Dec 92	30 Dec 92	28 Dec 92	30 Dec 92
01070	Abbreviations & Symbols	0.5	0	100	17 Dec 92	19 Dec 92	18 Dec 92	20 Dec 92
01080	Identification Systems	1	0	100	17 Dec 92	19 Dec 92	17 Dec 92	20 Dec 92
01090	Reference Standards	1	0	100	17 Dec 92	19 Dec 92	17 Dec 92	20 Dec 92
01100	Special Project Procedures	1	0	100	10 Dec 92	12 Dec 92	11 Dec 92	15 Dec 92
01200	Project Meetings	1	0	100	10 Dec 92	12 Dec 92	11 Dec 92	15 Dec 92
01300	Submittals	1	0	100	17 Dec 92	19 Dec 92	17 Dec 92	20 Dec 92
01400	Quality Control	1	0	100	10 Dec 92	12 Dec 92	11 Dec 92	15 Dec 92
01500	Construction Facilities & Temp Controls	1	0	100	17 Dec 92	19 Dec 92	17 Dec 92	20 Dec 92
01600	Material & Equipment	30	0	100	20 Dec 92	04 Jan 93	22 Jan 93	25 Jan 93
01650	Starting Of Systems/Commissioning	1	0	100	22 Jan 93	25 Jan 93	26 Jan 93	27 Jan 93
01700	Contract Closeout	1	0	100	28 Jan 93	29 Jan 93	29 Jan 93	30 Jan 93
01800	Maintenance	10	0	100	30 Jan 93	31 Jan 93	10 Feb 93	15 Feb 93
Subtotal:		**66**	**0**	**100%**	**70**	**78**	**30**	**35**
Phase 3:	**Site Work**							
02010	Subsurface Investigation	10	0	100	31 Jan 93	31 Jan 93	11 Feb 93	16 Feb 93
02050	Demolition	3	0	100	16 Feb 93	18 Feb 93	19 Feb 93	21 Feb 93
02100	Site Preparation	30	0	100	16 Feb 93	18 Feb 93	19 Feb 93	21 Feb 93
02140	Dewatering	6	2	80	20 Feb 93	22 Feb 93	25 Feb 93	27 Feb 93
02150	Shoring & Underpinning	3	0	100	25 Feb 93	27 Feb 93	27 Feb 93	30 Feb 93
02160	Excavation Support Systems	3	0	100	25 Feb 93	27 Feb 93	27 Feb 93	30 Feb 93
02170	Cofferdams							
02200	Earthwork	7	0	100	28 Feb 93	30 Feb 93	04 Mar 93	07 Mar 93
02300	Tunneling							
02350	Piles & Caissons	5	0	100	10 Mar 93	12 Mar 93	09 Mar 93	17 Mar 93
02450	Railroad Work							
02480	Marine Work							
02500	Paving & Surfacing	3	0	100	17 Mar 93	19 Mar 93	21 Mar 93	23 Mar 93
02600	Piped Utility Materials	2	0	100	17 Mar 93	19 Mar 93	21 Mar 93	23 Mar 93
02660	Water Distribution	3	0	100	17 Mar 93	19 Mar 93	21 Mar 93	23 Mar 93
02680	Fuel Distribution	3	0	100	17 Mar 93	19 Mar 93	21 Mar 93	23 Mar 93
02700	Sewer & Drainage	3	0	100	17 Mar 93	19 Mar 93	21 Mar 93	23 Mar 93
02760	Restoration Of Underground Pipelines	3	0	100	17 Mar 93	19 Mar 93	21 Mar 93	23 Mar 93
02770	Ponds & Reservoirs	3	0	100	17 Mar 93	19 Mar 93	21 Mar 93	23 Mar 93
02780	Power & Communications	3	0	100	17 Mar 93	19 Mar 93	21 Mar 93	23 Mar 93
02800	Site Improvements							
02900	Landscaping	20	0	100	17 Mar 93	19 Mar 93	21 Mar 93	23 Mar 93

Figure 7-1
Durations Factoring Fast-Track

These 37 activities are easily shown on the bar chart in Fig. 7-1, sequenced and computed for duration of activity originally estimated and actual duration remaining. Remember, this breakout of durations completion percentage represents only the first stage of a fast-track network. The second step is to put the phases of activities, along with the early and late start and finish events, into the formula.

The 37 activities can easily be scheduled with a manual bar chart or a timescaled logic diagram, produced on a PC with CPM software. With a pre-selected contracting plan developed by the project team to start preliminary activities prior to commencement, we get the combined elements of fast-tracking. Our CPM time-scaled network schedule now has the advantage of a head start prior to project commencement. The disadvantage is that the owner assumes most of the cost-performance risk of a cost-plus contract with prime contractors. But that risk management can be handled with a pre-selected contracting plan of a guaranteed cost & profit maximum plus incentive contract.

Once the durations percentages in Fig. 7-1 reach the thirty-four to forty-five percent completed design point, the contract could be let out to bid in a fast-track project execution. Now our schedule contains job logic, time manipulation tools of early and finish events, all the necessary elements of a velocity network diagram with activity interdependence linked by a timescale, plus the head start of contracted phased construction activities.

While the design/build fast-track CPM project execution will be in wide use throughout the construction industry for the next century, new variations of phased construction project scheduling are already becoming alternative construction management options. These include "just-in-time," process-oriented projects, and "bridging" project scheduling methods, which are variations of phased construction fast-track CPM, being pursued by developers, owners, architects, and engineers at an increasing rate.

Fast-Track CPM

The economic advantage of CPM fast-tracking is the cost control over the selected paths of critical activities running in series, and the many paths of activities running in parallel on the schedule. In addition, the early pre-commencement starts of the phased construction provide financial advantages to a project owner on a fast-track CPM project schedule, ideally reaching the earliest possible strategic project completion date. In fast-track CPM, the overall responsibility for controlling project execution remains not with a prime contractor, but with the owner. Typically, responsibilities for completion of the project design and specification are delegated to the project engineer, long-lead item procurement and construction activities to the project manager, supervision of field personnel to the site supervisor, and CPM schedule development and tracking to the project scheduler. Mutual cooperation between design professionals and the project team is essential throughout the project in this type of project execution. The project scheduler and prime contractor make their field schedules in close cooperation with each other. This early contractor cooperation with the project team produces a smooth and effective design-construction interface, and further ensures a viable and cost-effective CPM network schedule, thereby maximizing the potential for fast-track project profit. Phased construction projects create a combined-resource approach of an integrated owner/designer/construction team, which is why fast-tracking is becoming a major CPM project scheduling productivity enhancer.

One of the main differences in fast-track CPM, as opposed to standard CPM, is the handling of long-lead items. In fast-track CPM, the project scheduler accelerates the delivery of all long-lead items to the project site. These are now scheduled to intersect early construction phases at key activity starting events, thus accelerating the required long-lead items field delivery dates along with the activity's early starting event. Early quantitative survey material takeoffs also permit construction procurement personnel to order critical bulk materials in advance, such as lumber, pipe, and fixtures. Those activities are especially important in phased construction project execution.

In fast-tracking, preconstruction input by the subcontractors and project team begins while the architect is still in the design phase, to prevent design changes later in the schedule. Prime contractors then compensate for not having detailed finished set of plans and specifications to make their bids with cost-plus contracts tied to whatever actual costs are incurred. Activity subcontractors' bids are usually figured by square footage takeoffs, tied to whatever finished design costs are incurred regardless of projected costs. The main fast-track CPM advantage in project execution is speed of project completion. This contracting environment requires early, close, and formal coordination by the owner between the contractors, project team, and architect.

For a fast-track CPM schedule to be viable and operational, the plans and specifications need to be thirty to forty-five percent completed before the CPM project schedule is developed and executed. Because the project schedule must interlink with other major project considerations, such as design and long-lead items procurement, getting a fast-track's jump start without adequate finished design can incur costly errors. In fabrications with long-lead delivery times, error is obviously probable if the groundwork has not been done previously to provide accurate activities duration.

Cost-Loaded CPM

Cost-loaded CPM comes into use with unit-price contracts. These contracts use prices based on an itemized breakout of all work activities involved in the contract. Each is assigned a unit price. The final line-item price on each activity item is then computed by factoring the unit price times the item quantities. The unit price is then used as the controlling price for each item. Cost-loaded CPM is built from the unit-price bid sheets supplied in the original contract by the activities subcontractors.

Cost-loaded CPM is used by the scheduler as a schedule of values, and must not be confused with the lump-sum method. In a lump-sum project, the contractors contract to do all the work defined in the plans and specifications for a fixed price stated in the contract. Any quantity variations that are not added by the project team as change orders are not factored in. To cover these possible variations, a schedule of values is also requested from the contractors in their bids. A schedule of values is basically a unit-price bid with some notable differences. For example, item prices are shown for ease in making monthly progress payments only; they cannot be used for price change orders and imply no contractual obligations. Cost-loaded CPM is used to advance phased scheduling on progressive milestones.

A common problem that occurs from owners requesting cost-loaded CPM is their assumption that cost-loaded CPM will always result in easier progress payments management. This is true of lump-sum projects, but the situation is reversed in unit-price contract projects, because a cost-loaded network does not synchronize with the payline items on unit-price contracts. Therefore, in a lump-sum contract with cost-loaded CPM, the unit-price breakout should be submitted after the contract approval but before the first progress payment. In a lump-sum contract without cost-loaded CPM, a schedule of values should be submitted after the contract approval but before the commencement of work.

Typically, all engineering projects are based on a quantitative survey unit-price contract, in which the actual unit prices stated in the original bid will be held as the factor for all progress payment disbursement amounts. Because a unit-price contract does not have a fixed ceiling price, the final cost to the owner will be determined by the quantities actually completed. For the scheduler to adequately protect the financial investment of the owner, the scheduler must be careful to determine if the field quantities are precise. Conversely, in a lump-sum contract project, if an error is made within the first month of work it can be compensated for in a later progress payment. The contractor cannot bill for more than the stated fixed price, even if some of the final quantities varied from the anticipated amounts. In a lump-sum contract, contractors agree to complete project as per plans and specifications. Anything else necessary to complete the project must be considered as part of the contract, even if not specifically stated in the plans and specifications. Conversely, in a unit-price contract, the contractor may still be required to construct a complete functional project, but if any variation occurs in the quantities of any of the separate bid items listed, the contractor is entitled to an amount of money equal to the unit price of the bid item multiplied by the actual quantity of that item that was supplied or installed.

Chapter

8

Business Combat

Predictable Perils

Contractual financial penalties for late completion of a contractor's activity are liquidated damages that are daily cash *penalties*, which start in the thousands of dollars per day and go ballistic from there. On the other side of the coin, delays of the contractor's work schedule due to owner, architect, or project team fault are back-chargeable to the project in the form of *claims*. Lost time, lost project profits, lost use of the project in terms of unrecoverable capital income loss to the owner, liquidated damages charged to the general contractor, probable litigation among all parties, and poor quality workmanship are all results from overruns in network scheduling. Because of the amount of money involved, vested parties and contractors frequently run into expensive disputes. Usually, both parties throw away even more money by hiring lawyers and expert witnesses, and going into all-out litigation warfare.

These problems can usually be traced to inadequate float assigned within the critical path(s). This condition always burns up the allowable total float in the project at the first sign of trouble, and any further float necessary later on in the project must be "borrowed" from the remaining activities. I use the term "borrowed" because that is how it is defined in the trade. However, in reality, what does not exist cannot be borrowed. That's the truth of a CPM schedule, pure and simple (although the truth is rarely pure and never simple).

Disputes among the chain of command during the course of your project's production will be commonplace. Stuff happens all the time. Court dockets are full of construction litigation suits. Construction management professionals are currently hired by lawyer and legal referral systems at a prevailing rate of three to five hundred dollars per hour and upwards, plus all expenses, to appear as Expert Witnesses. In three of the cases I have sat in on, the project schedulers as well as the project managers were subpoenaed and/or depositioned. All but one of them were intimidated by the proceedings and made a weak showing of the strength of their CPM schedule's audit trail and accompanying back-up documentation.

Requiring the contractor to construct and maintain a CPM schedule has at least three advantages that should be utilized fully. First, a CPM schedule requires the contractor to work more efficiently. Second, it gives the owner notice of the actual progress of the work. Third, from a litigation standpoint, requiring the contractor to maintain a CPM schedule helps prove or disprove claims and helps to quantify the impact of an owner-caused delay. Many contractors do not usually have high-degree sophisticated software in their computer programs, and the odds are against a large CPM network being scheduled efficiently by a contractor showing ten different critical paths with the same duration. Invariably, this type of scenario results in unforeseen change orders and claims that typically go to litigation. This situation can often be foreseen in a contractor's network CPM schedule, while the same situation will not show up in project phase scheduling.

Claims

Contractors using their own CPM network scheduling can circumvent some of the general conditions and specifications of the contract by the use of claims. The contractor anticipates that the owner will delay at least two of the critical paths (on a multiple critical path network) that make up the project's production schedule, thus allowing the contractor to file claims against the owner. Claims are monetary reimbursements, usually in cash or cancelling off certain punch-list items chargeable to the contractor, that are levied against the owner for delaying the contractor's work schedule. This always results in cost overruns and blowing the budget. The tactic of eliminating as much float as possible from the network puts the owner at a cost disadvantage in network analysis, and smart contractors know this.

So, the experienced owner may try to nullify this contractors' claims technique through clauses in the schedule specification that direct the contractor to redraw the network any time it is behind schedule on any specified critical path. In the accompanying CPM software, this condition is instantly shown on the Sort by Total Float file in the Negative Float column whenever an activity falls behind schedule. The owner's project team will attempt to resolve these and other scheduling problems, and will always move to gain management control over claims. Usually, the project manager will make the stipulation in the contract for one or more of the several network scheduling techniques currently available. Such specifications would be written with the intent to boilerplate (cover all bases for) the owner's protection. With a combination of appropriate general conditions and supplemental general conditions, such specifications can be quite complex. One can easily see that the whole process can, and often does, wind up in claims and litigation. Many court decisions have been rendered on such claims, but none seems to have solved the problem. Now that you see both perspectives, you can understand how, as a project scheduler, your use (or non-use) of different types of float in your phase scheduling can benefit either the owner (if that is your client) or the contractor (if that is your client).

This is how we do it. To prevent creation of artificial networks by the contractor, those schedules traversed by more than one critical path should require a contractual agreement between the owner and the contractor prior to the start of the work or any change in work's scope. To prevent manipulation of activity duration by the owner, networks with several near-critical paths should be subject to resource (labor hours, equipment hours, dollars, etc.) and quantity (permanent material) analysis with respect to activity duration. Owners and contractors should be required to review and approve such analyses prior to the start of work or start of change-of-scope work. In today's business combat, control in this area is another profit maker for the hot scheduler.

Such performance analyses should consider the following:

- Budgeted rate of quantity installation as a function of time.
- Budgeted rate of resource usage as a function of time. This can be done by resource-scheduling techniques currently available in most CPM processors.
- Budgeted consumption of resource usage as a function of quantity, type, and time.

The results of these calculations (or other performance calculations) should be agreed upon by all parties to each activity under potential dispute prior to performing the work. To prevent improper consumption of float by either party, distributed float calculations can be used. Distributed float allocates float to each and every activity. Using such a targeted approach will always predetermine who gets what float. Relative to whom has hired you, you now know how to swing the scales on their behalf.

AIA Document A201, Section 4.3.1, delineates the parameters of acceptable and enforceable delay claims submitted by the contractor. These requirements can make it difficult for a contractor who does not keep detailed weather records to claim a time extension for adverse weather conditions. They reflect a belief by the AIA that weather generally is a risk assumed by the contractor, and that only in extraordinary circumstances should weather be the basis for a time extension. The legal documents used by the courts to determine judgments in weather related claims are a Daily Inspection Report (DIR) and a Quality Assurance Report (QAR). If the contractor or owner is unaware of these documents, then certainly you, the professional project scheduler, should be. And you should implement them on your client's behalf, whomever that may be, to protect your schedule.

Another reason for not granting a time extension is the single contract system's objective of centralizing administration and responsibility in the prime contractor. Only if subcontractor-caused delay is specifically included should it excuse the prime contractor. This needs to be stated within the contract specifications to protect your client from the schedule's default through circumstances beyond your control. The independent contractor rule, though subject to many exceptions, relieves the employer of an independent contractor for the losses wrongfully caused by the latter.

Prime contractors sometimes assert that the subcontractor is an independent contractor, inasmuch as the subcontractor is usually an independent business entity and can, to a large extent, control the details of how the work is performed. Even so, the independent contractor rule does not relieve the employer of an independent contractor when an independent contractor has been hired to perform a contract obligation, and the party who suffers the loss caused by the independent contractor is whom the contract obligation is owed. In the construction contract context, the owner usually permits the prime contractor to perform obligations through subcontractors. This does not usually mean that the prime contractor is relieved of its obligation to the owner, unless the owner specifically agrees to exonerate the prime contractor. The residuary power of agency granted to the architect to grant time extensions has been criticized as leading to a deterioration of any fixed completion date. It does have the virtue of not forcing the architect to think of every possible event and include it in the contract specifications as a "catalog of events" justifying a time extension.

Another portion of contract law affects the project schedule, and that is AIA Doc A201, Sections 4.3.3 and 4.3.8.1, which cover time extensions in construction contracts. These usually provide a mechanism under which the contractor will receive a time extension if it is delayed by the owner, or by designated events such as those described in the contract specifications. Increasingly, contractors make large claims for delay damages. As a result, it is becoming even more common for clauses in the contracts, particularly public contracts and private ones drafted with the interests of the owner in mind, to attempt to make the contractor assume the risk of owner-caused delay.

Public entities are often limited by appropriations and bond issues. As a result, they must know in advance the ultimate cost of a construction project. Thus, many public entities use the disclaimer system for unforeseen subsurface conditions. Similarly, they wish to avoid having claims made at the end of the project, based on allegations that they have delayed completion or required the contractor to perform its work out-of-sequence. These public works and awarding authority entities recognize that barring claims may cause higher bids, but they would prefer to see bidders increase their bids to take this risk into account rather than to face claims at the end of the job.

Looking at the troublesome question of the validity of no-damage clauses is not within the parameters of the project scheduler, but lies instead with the owner and/or the project team, so we will not delve further into this issue other than to make you aware of alternatives so that you, as a professional project scheduler, will have viable options to propose if you should so wish. One method used to eliminate or reduce the likelihood of such claims and delays to your schedule is to have the contract specify that the owner has the right to delay the contractor, and that such interference is not a contract breach. Another technique is to contend that any time-extension mechanism is the exclusive remedy. Generally, the availability and use of the time-extension mechanism does not imply precluded delay damages. The AIA's time-extension mechanism specifically states that it does not bar the contractor from recovering delay damages.

An indirect technique for limiting delay claims is to require written notice by the contractor, to the owner or the project team, if events have occurred that later will be asserted as justifying delay damages. This is the method that is currently taking preference nationwide in the construction industry, because it warns of potential claims in time to resolve the situation at the activity stage. Typically, this notice is stated to be a condition precedent to any right to delay damages. Historically, although such clauses have not been looked on favorably by the courts, noncompliance with a notice provision can be the basis for barring a claim for delay damages and as a contractual obligation, can "bulletproof" the situation as well as is currently possible without undue and unjustifiable constraint on the contractor.

Another common method to deny the delay damage problem outright by not setting up notice conditions or specifying what can be recovered, is to configure the contract with no-damage or no-pay-for-delay clauses. Such clauses attempt to place the entire risk for delay damages on the contractor and to limit the contractor to time extensions. Generally, such clauses are upheld but are not looked on favorably by the courts and much less by contractors. A current modification on the "no-pay"or "no-damage" provision clauses is to provide that the contractor can recover delay damages only after a designated number of days' delay by the owner.

A newer technique is to have contract clauses that permit some delay damages but avoid large claims for diminished productivity by an open-ended, total-cost formula. The method provides that the contractor can recover delay damages, but only for premiums paid on its bond and for wages and salaries of workers needed to maintain the work, the plant, and the equipment during the delay. California courts, however, have held that such clauses will not be enforced if there is affirmative or positive interference or a failure to act in some manner essential to the prosecution of the work. These clauses have functions and ramifications that go deeper, and are beyond the scope of this book, and will be found in greater detail in a future reference book on Construction Contract Law, by me. Such a reference book should be included on every project scheduler's bookshelf as protection against schedule delays beyond the scheduler's immediate ability to control and manage. Whether such clauses should be included in a contract depends both on the likelihood of their being enforced in a proper case, and on the willingness of the owner to take "front-loaded" costs (at the front end of the contract) rather than through claims at the end of the work. However, the use of these clauses in a highly competitive construction market will protect the project schedule somewhat by encouraging bidders to bid low and take their chances.

Dispute Resolution

Velocity network diagram scheduling systems were originally developed as production management and cost-control tools in Europe, but like everything else efficient at making money, velocity network diagram scheduling has evolved significantly. Schedulers learned how the system could be abused for their own advantage. Some developers and project teams have even used CPM scheduling as a litigation weapon, to secure as much profit as possible through denial of claims by contractors.

Conversely, prime contractors and subcontractors have been known to employ professional project schedulers to devise ways to give their particular business an edge on float in activity events, especially those on critical path(s). Modern computer network-based management systems that run a construction project like a mathematical clock can also create biased documents to support the originator's right to claims and extra work orders. Those who are familiar with software spreadsheets will nod in agreement with me here as I inform you that anyone can make wrong data look right in charts if it's presented with enough spreadsheet manipulation and corresponding bar graphs.

On one hand, a contractor might create an artificial network with multiple critical paths. The intent of the contractor would be to present claims if the owner caused delay on any of the paths. On the other hand, the owner might plan the project duration and then shorten it. The owner's intent would be to obtain a bid on the shortened duration then hold the contractor to the time limit. Obviously, these practices reduce an otherwise effective management tool to a weapon for justifying or denying claims. The result has been more litigation over more claims.

Contractors and owners view the problem differently. A contractor might question whether an owner has a right to direct starts or delays for specific purposes. Owners, trying to manage project costs and overall scheduling, might ask whether they have the right to force the contractor to start an activity before an established late-start date. Too many of these late starts could cause problems toward the end of the job. For the owner, a week's slippage in a power plant's commercial operation date translates into a loss of several hundred thousand dollars. At the same time, improper or premature activity starts can cause serious hardship for a contractor, who may incur additional expense for equipment, material, labor, and other resources.

Very little float is in a network velocity diagram, and the owner is subject to claims at any time that it delays the prime contractor on any one of the multiple critical paths. This technique invariably gets the prime contractor off on the wrong foot when he submits the claims. The owner's project team will then resent the owner's implication that they were not smart enough to see through the contractor's ploy, and an adversarial relationship will be set up that will last for the duration of the contract. When requiring a CPM schedule submittal from a contractor, it is wise to follow the Associated General Contractors guide specifications, which recommend that the network be comprised of individual work activities, each of which does not exceed ten days' duration.

However, remember, information is power. These tactics are the domain of the professional project scheduler and professional trade secrets should not be confided to others on less than a need-to-know basis. As Shakespeare wrote, once the deed is done, 'tis done forever and cannot be undone. Your prowess in these matters may or may not be something you wish others to know about. It's therefore recommended that your knowledge of manipulation techniques in network velocity diagramming be kept close to your vest.

Here's how we do it. Dispute resolution should be a contractual obligation appearing within the specifications. The order of precedence of resolution steps should be: mediation, arbitration, litigation... specifically, and in that order. Arbitration should be further specified as binding arbitration. In this manner, precedence is already established for resolving the inevitable arguments that will occur throughout the project's production.

Mediation

Mediation means that all parties try to iron it out amongst themselves, with no outside consultants or lawyers becoming involved. If the contract is written with the above precedence for dispute resolution, and if the dispute cannot be resolved at the mediation level, all parties must proceed to arbitration before leaping into litigation.

Arbitration

Arbitration proceedings are much more informal than court litigation proceedings. Arbitrations are generally held in conference rooms in private locations. Arbitration is usually heard before a retired judge or member of the Bar Association, and no lawyers are allowed. Arbitration is a lot less expensive than going to court. It is also a means to a quicker solution. I have always advised my clients to include a binding arbitration clause in all contracts. If this clause is included, all parties are forced to arbitrate. The downside to arbitration is that, although arbitration is cheaper and quicker than court litigation, if your side loses it is almost impossible to appeal the judgment decision in arbitration. Proceedings are arranged and conducted by the American Arbitration Association. However, the actual proceedings are left to the arbitrator's discretion. Meeting considerations, such as time, location, and agenda, are worked out among both parties and the arbitrator.

Arbitration expenses can sometimes be substantial, although typically less so than litigation. In arbitration cases lasting only one or two days, the arbitrator charges no fee. On the third day, the arbitrator begins to charge a fee that will be comparable to other legal professionals in the area (which will usually average three digits an hour), and proceedings will begin to get costly. Additionally, there is a filing fee for arbitration that is based on the size of the claim. This is usually recoverable from the losing party by the winning party. These fees can be split between both parties as part of the final award, at the discretion of the arbitrator.

Arbitration is typically faster than a court suit. Court litigation (especially in California) usually takes up to a year and a half to get to the first court date. The arbitration process, which includes the filing process, arbitrator selection, and calendar date, usually takes only a month to schedule. The entire arbitration settlement can take up to six months, typically one-third of the time required for litigation cases.

Litigation

Nobody wins but the lawyers.

Typical Case History

A contractor hired a professional scheduler and, using the scheduler's CPM network, constructed a $3 million commercial building exactly per approved plans, only to find out during inspection that the structure did not meet code compliance due to an error by the building inspector. Who was liable? This case scenario appears in many stories about planning or building departments being liable for mistakes they have caused or overlooked. Here are some facts based on previous court cases. Let's first deal with building officials. In several cases, such as Chaplis v. County of Monterey, the court held that, while the building official issued a permit in violation of a county ordinance, the county and its employees were immune from liability. However, when the trial court in a construction-defect case dismissed a complaint against a building department official in Mammoth Lakes, the appellate court reversed the decision and held that the plaintiff stated a good cause if the representations made by the building official were shown to be false. So you see, it could go either way. Courts have traditionally held that building departments and their employees are immune from liability, and that errors in design that are carried through the construction process revert to the architect or design professionals who are held to be liable, as they are supposed to be experts and are held to a "higher level of responsibility and accountability," and should be informed enough to spot an error before or during construction.

A subcontractor on a $5 million civil construction project submitted a CPM network schedule that had been meticulously prepared. Its court deposition stated that the only objective of the CPM schedule was to get the project done as quickly and cost-effectively as possible. Due to a public works department-caused delay, the subcontractor fell behind schedule. Later, the subcontractor submitted a claim for additional money and a 45-day extension. The awarding authority's engineers analyzed the prime contractor's network and showed that subcontractor's succeeding activity event on the next major milestone on the critical path was delayed only 20 days. The awarding authority contended that 20 days' delay was all the subcontractor was entitled to. The subcontractor stated it was only through its diligence in accelerating its work that the delay was held to only 20 days, and the subcontractor should be entitled to the full 45 days.

The final court judgement was that the subcontractor could not support its 45-day claim because of improper schedule monitoring. The awarding authority's engineers had been able to do a better job of network analysis than had the prime contractor, and the subcontractor had to settle for the expenses he could prove. Despite the fact that it was public works' delay, the subcontractor was, in effect, financially penalized for maintaining the schedule because of its activity's improper schedule monitoring by public works.

In Miller v. Delcon Development Corporation, Mr. Miller was a subcontractor who complained to the project owner, Delcon Corp, that the prime contractor had failed to pay him for his time and materials. The owner wrote a letter to Mr. Miller requesting that Miller send them a stop notice. Mr. Miller never did, but he did file a Preliminary Notice in a timely manner. The court held that if a subcontractor in a CPM network contract does not respond to a letter such as the one written by the owner, he loses his lien rights and stop notice rights. The final judgement was that Mr. Miller had lost his lien rights as a result of not giving the owner a stop notice. His only recourse was to sue for liquidated damages in civil court, which he tried and lost.

Multiple Critical Path

A prime contractor on a $14-million industrial building project submitted a CPM network in compliance with the owner's specifications. There were four zero-float paths and seven critical paths. The prime contractor later stated that he was anticipating that the project team would delay one or more of the seven activities that were a part of the critical paths, thus allowing the contractor to file claims against the owner, which the contacting company did at project completion. This type of bid getting is known in the trades as "low-balling." During mediation, the owner's project team cited the seven critical paths as an unrealistic approach. The contractor responded that it was the contractor's plan of work and the owner did not have any right to alter it once the contract had been signed. The mediation went unresolved into arbitration and later into litigation.

What had the contractor done to his network, and what was achieved? First, he increased the duration of routine concrete pours to three times their normal duration. If requested to shorten this duration, he planned to request payment for accelerated work. To prevent this, the owner had, as stated in the contract specifications, the right to use CPM and other standard planning tools, such as estimating guidebooks, to determine reasonable durations of all activities. Exceptions were to be negotiated between them both. The contractor used "policy or management constraints" (preferred way of doing work) to consume float. These constraints prevent activities from starting until preceding activities are sufficiently complete. On detailed analysis, this appeared illogical; however, the contractor was asked to apply workaround techniques, and he countered by saying that the changes to his plan of work would result in extra work, and therefore more claims against the owner.

An owner should have the right to demand that constraints of this type be removed, and the work be replanned when the constraints threatened the critical path. Difficulties such as those just described might be overcome with adequate specifications. However, one can hardly hope to anticipate all the problems that can arise, particularly if the specifications writer has a limited knowledge of all the tricks that can be used.

The specifications should protect both the owner and the contractor. But it should be remembered that the owner pays for the scheduling system and is entitled to get what he pays for. The specifications should set forth restrictions on falsifying networks to eliminate float. They should spell out what rights the owner has to utilize float to his advantage. They should clarify areas where the owner has the right to apply standard estimating techniques to activities for which contractors have obviously set over-long durations to eliminate float.

Contract Change Orders

A contract change order is an authorization of additional work in addition to, but pertaining to, the original contract. The courts have typically held that a change order is a contract in itself, and since it amends the original contract, it must be reflected in the network schedule and filed with the original contract documents. The legal sequence that must be observed regarding change orders is that, if the network schedule requires numerous changes during the course of the project, the contractors must bill the owner after the changes have been completed. This is important to know, because to do otherwise is to allow your CPM network schedule to become sidetracked or derailed. If you, through the project manager, have to play hardball with the subcontractors to keep your schedule on track, so be it. The law says that a contractor may collect (in progress payments) only what he has provided in labor and materials to date. If, on Monday, the contractor performs additional work that amounts to $5,000, the contractor can amend (through a contract change order) the next progress payment to include that amount. However, the work must be done first.

Contractors have a right to collect extras as long as they amend their contract and/or use change orders indicating when extras are being added and when they are to be paid for. This means that activities cannot be suspended by the contractors for extra work payments, and that your schedule can not be brought to a halt over contract change orders. The primary objective of any project schedule is to complete the project as designed, in a systematic, coordinated manner. All this must be accomplished in the shortest time consistent with material and personnel constraints, thereby maintaining a good profit. Today's construction industry management people recognize that change orders are a normal part of the construction process. Every project manager, site superintendent, and project developer knows that, in all probability, there will be changes occurring on the project that will affect the final duration. If they can be handled competently and in stride, the effects on the project might still be felt, but accountability for them will remain where it belongs. Project schedules that are constructed and updated in a clear, consistent manner with complete, thorough notes and references can accomplish this. If the schedules are managed so that the information in them can be easily extrapolated, categorized, and supported (or denied) in the event of a worst case scenario, their strength will actually minimize serious disputes. A good schedule will easily provide significant, visible comparisons and convincing proof of damages, including delay, acceleration, suspension of work, inefficiencies, disruption, interference, and demobilization.

In dealing with change orders in your project schedule, remember, contract law does not allow the prime contractor or subcontractors to make any changes to an existing contract. An addendum in specified work requires the contractor to write up a contract change order and then seek written approval from the owner. The owner is not obligated to sign a change order, but the owner's approval is absolutely necessary to avoid litigation problems. The owner cannot be forced to pay for work that was not authorized by the owner. Contract law specifies that "all changes to the original contract must be in writing and signed by both the property owner and contractor."

These following six fundamentals form the basis for presentable evidence in a court of law for CPM schedules and will, in turn, establish the basis of good recordation, clear accountability, and effective presentation:

1. The CPM schedule must be the same one that was actually used to build the project. Even if a schedule that is substantially different from the one under consideration had been formally submitted and approved by the owner and/or the design professional, the schedule that was actually used and depended upon by the various trades will generally be considered to be the contract document.

2. The CPM schedule must be periodically revised. It is generally recognized that project scheduling, however exact or inexact, is nothing more than a plan to construct what can reasonably be construed from the contract documents, considering that there may be several ways to build the same project. Although contingencies may be included to allow for imprecisely defined variables, there probably never has been a schedule that anticipated every problem, coordinated every piece of work, and required no modifications to make it work. Changes to, or corrections of, the plan are inevitable. The schedule, therefore, must be updated periodically to maintain a current and accurate representation of reality.

3. The periodic updates of the CPM schedule must show all positive and negative influences by all parties. None must be singled out; none must be absent for convenience. If all updates, for example, only indicate delays caused by the owner and fail to delineate other known problems caused by other parties, it won't be very difficult for your opponent to demonstrate that bias has been built into the document. Failing to recognize all significant events will bring into question the validity of the entire CPM schedule presentation.

4. The CPM schedule must include realistic construction logic and activity durations to demonstrate the professional ability of the planning and scheduling team. Illogical sequences or the lack of consideration of critical variables (such as indicating the installation of commercial HVAC units in a building with no overhead doors, to occur before the roof is constructed) will only demonstrate that the constraints in the CPM schedule were unrealistic to begin with.

5. The CPM schedule must fairly represent the actual method planned to build the project. If, for example, the CPM schedule had been prepared primarily to cater to progress payments, or was otherwise unrelated to actual construction of the project, it would become clear that, as a tool for managing the project, its value was marginal at best.

6) The CPM schedule updates and analyses must be realistic and in perspective. An overly aggressive computation of damages might hurt the validity of the entire analysis. Direct cause-effect relationships must be shown.

In addition to the foregoing, the project scheduler must be sure that all notes and references are correlated with the more detailed, chronological project correspondence. Have all field notes, claims, commitments, deadlines, and promises detailed in the schedule documents. Include the dates and the names of the individuals in those organizations supplying information. Pin it down. I wasn't kidding when I said litigation warfare.

Chapter

9

Contract Law

Legal Aspects of the Project Scheduler

Contract law is for lawyers, so why on earth is it included in a book on project scheduling? For the same reason I have included chapters on construction law and estimating. All three of these disciplines directly influence and exert controlling force, for better and for worse, on portions of your CPM network schedule. I would be negligent as your instructor if I did not point that out and give you the basics of how each affect the success (or failure) of your CPM production schedule. All of these next three chapters, on contract law, construction law, and estimating, will have an influence on critical duration of your project and, hence, your project schedule. They are all big enough areas to write separate books on, so accordingly, only the basics of each that are relative to CPM network scheduling will be covered in this edition.

The courts historically have not used consideration of time in CPM schedules as affecting judgments in construction litigation lawsuits. This is due to the last decade's tendency for frequently delayed construction projects to end up in litigation. Timely completion depends upon proper performance by the many participants, as well as optimal conditions for performance, such as weather. Nowadays, encountering unanticipated subsurface conditions is the kiss of death. Inspections, excavations, and soil remediations are just a few of the costly delays that can result. Delayed performance, due to encountering unanticipated subsurface conditions, has been held by the courts to be invalid grounds for termination of the contract. This means the owner and project team continue to eat it financially. Delayed performance is also unlikely to create legal justification for the owner's refusal to pay the contractual compensation. At this point, the owner is typically left with inadequate damage recovery. Insurance policies, and bonds to protect damage recovery in such cases, are extremely expensive and typically put the project out-of-budget and therefore make it not viable.

Similarly, delayed payment by the owner is less likely to automatically give the contractor a right to stop the work or terminate its performance. Finally, a strong possibility exists that a performance bond that does not expressly speak of delay will not cover delay damages. These damages will then be contested, and it falls to the scheduler to prove the validity and logic of the project schedule.

Computation of delay damages is also different. If the owner does not pay or the contractor does not build properly, measuring the value of the claim is, relatively speaking, simple. If the owner does not pay, at the very least, contractors are entitled to interest. If the contractor does not build properly, the owner is entitled to the costs of correction or the diminished value of the project.

Delay creates serious measurement problems in CPM schedules. The owner's basic measure of recovery for unexcused contractor delay is lost use of the project. The contractor's basic measure of recovery for owner-caused delay is added expense. Lost use is difficult to establish in noncommercial projects. Added expense is even more difficult to measure. Because of measurement problems, each contracting party, whether it pictures itself the potential claimant or the party against whom the claim will be made, would like a contractual method to deal with delay claims, either to limit them or to agree in advance on amounts. This does not mean that time is not important in construction. The desire to speed up completion, due to money investment pressures, is what employs the professional project scheduler.

Traditionally, two schedules are initially purposed for any project. The general contractor supplies a version suited to his company's timing for profit, and the owner supplies one suited to his timing for profit. The key factors for these two schedules to merge successfully is how close each is to the critical design, procurement, and construction of all aspects of the project. Once the schedule is completed, it needs to be cooked. This term means a critical analysis of the schedule by the project scheduler and the project manager (also called the construction manager). At this point, the schedule is debugged by using "what if" scenarios through the critical paths of production. These first editions of the schedule are generalized overlays, called macros. Macros are then redesigned to reflect sequential activities' interdependencies and events timescaling. Important events and individual phase completions are noted as milestones.

Owners typically tend to be overly optimistic about all activities happening with no problems, their schedules showing an unrealistic and often unachievable project completion date. On the owner's schedule, activities are strung together in a tight series, with each activity closing event also being listed as the starting event for the next activity. In this type of configuration, all activities lie on one critical path. This makes for a network diagram with little or no flexibility in scheduling float allowances, which is the primary area a scheduler makes adjustments in. General contractor's schedules, conversely, show free float time in their subcontractors' related activities to serve their own work schedule, usually with no free float linkage to overall project total float. Obviously, with what you have already learned about the vital management uses of float, you can see that neither of these two schedules will be adequate.

What invariably happens in a real-world field application is that the initial activities burn up all available total float, and any emergency float time is then borrowed from the succeeding activities, leaving little or no float time along the critical path. This always results in the critical condition of production and cost overruns, because contractors do not want to lose the bid and will often accept an unrealistic deadline, hoping that they can pull a rabbit out of the hat somewhere along the line.

AIA Document A201, Section 3.10.1 requires the contractor to submit its construction schedule for the information of the owner and the architect. The schedule must provide "for expeditious and practicable execution of the Work." The contractor's failure to conform to the most recent schedule constitutes a breach of contract under Section 3.10.3. Contracts prepared by experienced public or private owners, particularly private owners under the influence of their lenders, usually prescribe much greater detail and take greater control over the contractor's schedule.

This can manifest itself in language requiring that the CPM schedule be on a form approved by the owner or the owner's lender; that each monthly schedule specify whether the project is on schedule (and if not, the reasons therefore); that monthly schedule reports include a complete list of suppliers and fabricators, the items that they will furnish, the time required for fabrication, and scheduled delivery dates for all suppliers; and that the contractor hold weekly progress meetings and report in detail as to schedule compliance.

Similarly, the Engineers Joint Contract Documents Committee (EJCDC) takes progress much more seriously than does the AIA. For example, the EJCDC's number 1910-8, Section 2.6, requires the contractor to submit, within ten days after the effective date of the contract, an estimated progress schedule, a preliminary schedule of values, and a preliminary schedule of submittals. The finalized schedule must be acceptable to the engineer, and Section 6.6 requires the contractor to submit, for acceptance, adjusted progress schedules. The approved finalized schedule becomes a contract document.

Outside Delays

Referring to the software program, you will find that the field daily reports have columns for weather delays. An inclement weather system passing through your area that turns out to be a toad floater will also delay outside activities, such as excavation, grading, paving, and placement of concrete. These are considered constraints of practical limitations and must be factored as such in the CPM schedule. The reasons are twofold. First, time cushions must be installed within total float, to provide allowances in practical production parameters. And second, AIA Doc A201, Section 4.3.8.2 (which is the section dealing with claims) requires the contractor to document "by data substantiating that weather conditions were abnormal for the period of time and could not be reasonably anticipated" before the contractor can receive a time extension for weather conditions. Section 4.3.8.2 further requires that the contractor document that the weather conditions "had an adverse effect on the scheduled construction."

The overall precedence concerning CPM schedule delay is found in AIA Doc A201 Section 8.3.1, which states: "If the Contractor is delayed at any time in the progress of the Work by any act or neglect of the Owner or the Architect, of an employee of either, or of a separate contractor employed by the Owner, or by changes ordered in the work, or by labor disputes, fire, unusual delay in deliveries, unavoidable casualties, or any causes beyond the Contractor's control, or by delay authorized by the Owner pending arbitration, or by other causes which the Architect determines may justify delay, then the Contract Time shall be extended by Change Order for such reasonable time as the Architect may determine."

This section delineates the parameters of acceptable and enforceable delay claims submitted by the contractor. These requirements can make it difficult for a contractor who does not keep detailed weather records to claim a time extension for adverse weather conditions. They reflect a belief by the AIA that weather generally is a risk assumed by the contractor, and that only in extraordinary circumstances should weather be the basis for a time extension. The legal documents used by the courts to determine judgments in weather related claims are a Daily Inspection Report (DIR) and a Quality Assurance Report (QAR). If the contractor or owner is unaware of these documents, then certainly you, the professional project scheduler, should be. And you should implement them on your client's behalf to protect your schedule.

Another reason for not granting a time extension is the single contract system's objective of centralizing administration and responsibility in the prime contractor. Only if subcontractor-caused delay is specifically included should it excuse the prime contractor. This needs to be stated within the contract specifications to protect your client from the schedule's default through circumstances beyond your control. The independent contractor rule, though subject to many exceptions, relieves the employer of an independent contractor for the losses wrongfully caused by the latter.

As noted previously, prime contractors assert that subcontractors are independent contractors because the subcontractor is usually an independent business entity and can control the details of how that activity is performed. Even so, the independent contractor rule does not relieve the employer of an independent contractor when the independent contractor has been hired to perform a contract obligation and, under that employer's direction, the party suffering the loss caused by the independent contractor is the party to whom the contract obligation was owed.

Another portion of contact law affects the CPM project schedule, and that is AIA Doc A201, Sections 4.3.3 and 4.3.8.1, which cover time extensions in construction contracts. These usually provide a mechanism under which the contractor will receive a time extension if the project is delayed by the owner or by designated events such as those described in the contract specifications. Increasingly, contractors make large claims for delay damages. As a result, it is becoming even more common for clauses in public works contracts to attempt to make the contractor assume the risk of owner-caused delay. Public entities are limited by appropriations and bond issues, so they must contract in advance for the full cost of the project. To do this, many public entities use the disclaimer system for unforeseen subsurface conditions. Similarly, they wish to avoid facing claims at the end of the project, based on allegations that they have delayed completion or required the contractor to perform its work out of sequence. Repeated here for ease of reference are the management techniques for limiting delay claims in a CPM schedule.

The *first* management technique is to require written notice by the contractor to the owner or project team if any events have occurred that later will be asserted as justifying delay damages. This is the method that is currently taking preference in the industry, as it provides warning of potential claims in time for someone to do something about the situation in the mediation-dispute-resolution stage, and resolve it at the activity level. Typically, this notice is stated to be a condition precedent to any right to delay damages. Historically, although such clauses have not been looked on favorably by the courts, noncompliance with a notice provision can be the basis for barring a claim for delay damages, and because it's a contractual obligation, can "bulletproof" the situation as well as is currently possible without undue and unjustifiable constraint on the contractor.

The *second* management technique is to deny the delay damage problem outright by not setting up notice conditions or specifying what can be recovered. The method is to configure the contract with no-damage or no-pay-for-delay clauses. Such clauses attempt to place the entire risk for delay damages on the contractor, and to limit the contractor to time extensions. Generally, such clauses are upheld but not looked on favorably by the courts and much less by contractors. A current modification on the "no-pay"or "no-damage" provision clauses is to provide that the contractor can recover delay damages only after a designated number of days' delay by the owner.

Legal Notices

During the project's CPM schedule timeline, certain legal notices must be filed and tracked. The primary reason for this is litigation protection from nonpayment of services. When an activity subcontractor is not paid by your project's prime contractor, that subcontractor can file a Mechanic's Lien and/or Stop Notice at any time after ceasing to perform labor or furnishing materials, and until thirty days following the Notice of Completion or issuance of a Notice of Cessation. A prime contractor, on the other hand, if not paid, can file a Mechanic's Lien within sixty days after Notice of Completion or issuance of a Notice of Cessation. Let's look at each legal notice closely:

Notice of Completion

Notice of Completion is normally filed by the project owner within ten days after the project is completed. This legal notice restricts the number of days within which a contractor can file a lien against the property. This should also equate to a milestone in your schedule that marks the ending event of the project production and the starting event on project closeout. Once this is filed, contractors should do no more work. The courts have held the criteria to determine the completion of work to be:

1. The occupation or use of a work of improvement by the owner.
2. The acceptance of work by the owner.
3. If there has been a cessation (stoppage) of labor on the job for a period of 60 days without a notice of any kind being recorded which constitutes a legal completion.

Notice of Cessation

This is the legal notice used if, for any reason, the job has been stopped because of a union strike, trucking strike, labor dispute, labor shortage, or a similar reason. It cannot be recorded until there has been a cessation of labor for thirty days, with the exception of Force Majeure.

Notice of Non-Responsibility

This notice is filed by a project owner in a conspicuous place upon the property within ten days after obtaining knowledge that work that was not ordered is being done. This notice must also be recorded with the county recorder in the area of jurisdiction wherein the property is located, giving notice that the owner will not be liable for any accident or injury occurring.

Preliminary Notice

Preliminary Notice is a contract document filed by subcontractors and suppliers within twenty days of beginning their activity on the project. On residential and commercial jobs, Preliminary Notice is sent to the lender, the owner, and the prime contractor by registered or certified mail, or is delivered in person. On public jobs, Notices are sent to the awarding authority representing the public agency and the prime contractor. Preliminary Notices are filed to establish the subcontractors' and suppliers' rights of lien and stop notice. Although no company may file a lien on a public works job, because the property in question is public property and cannot be liened, subcontractors and suppliers on public works jobs do have stop notice rights in the event of delay in progress payments. However, if Preliminary Notice is filed late on a public works job, all stop notice rights of the contractor are lost.

Preliminary Notice must be filed by subcontractors and suppliers within twenty days of their beginning work on their specific activity within the CPM project schedule. Therefore, it is mandatory that every licensed contractor doing any activity within your project schedule, with the exception of the prime contractor, give its notice not later than twenty days after it has first furnished labor, services, equipment, or materials to the project's jobsite. The prime contractor does not give a Preliminary Notice, due to the fact that its contract with the project owner is the first, or prime, contract and serves as Preliminary Notice.

The purpose of Preliminary Notice is to notify those vested parties of the project who never have a direct relationship with (and often never see) the activities subcontractors who do the work. One of your bar sorts should track submission of these legal notices, and one of your timeline sorts should track their timetables. Contract law statutes require that if any work is done by any of your project's subcontractors for over $400, the subcontractor is subject to disciplinary action by the Contractors License Board for not filing this legal notice. If notice is given later than twenty days on a private job, the contractor may not recover the value of work done at more than twenty days of his activity construction from the date of the notice. Lien rights are lost as to anything furnished more than twenty days before serving of the notice.

An activity subcontractor not having a direct contact with the prime contractor, other than a laborer, must, as a necessary first step to the validity of his Stop Notice rights, give a written notice to the contractor and the public body (awarding authority) concerned, within twenty days after he has first furnished labor, services, equipment, or materials to the jobsite. This notice takes the form of Preliminary Notice.

Contrary to the law on private jobs, if the claimant on a public works job lets his first twenty days of activity work elapse without filing Preliminary Notice, said contractor loses all stop notice rights forever. But that contractor still has a legal right of action and can sue against the bond and the bonding company, since all public works jobs are bonded. These public works bonds protect all those involved in the project construction from non-payment for their services.

A 90-Day Public Works Preliminary Notice is necessary to enforce a claim on a payment bond on public works jobs. It is required in public works projects by subcontractors and suppliers to the prime contractor, within ninety days after furnishing labor, materials, or services.

Mechanic's Lien

Mechanic's Lien is a lien upon the project's real estate to secure the compensation of those who have been directly instrumental in the improvement of such property. This is a critical factor of final payment for the services of a professional scheduler in that, should the project go over budget and contractors or materials suppliers not be paid in a progress payment, mechanic's liens filed may halt the production process. The project may wind up in litigation and your invoices for your services may go unpaid. In such a case, if you are an outside-services contracted scheduling consultant, you may need to file a mechanic's lien to secure payment for your services.

Wage earners do not have lien rights, but materials suppliers, contractors, and laborers have a lien upon the property upon which they have bestowed labor or furnished material for the value of such labor done and materials furnished. The legislature provides, by law, for the speedy and efficient enforcement of such liens. Mechanic's lien law is of particular interest to all persons in the building and construction industry, inasmuch as the value of labor performed and materials furnished to any building or construction project are debts, the payment of which can be secured by a lien binding the real property as security for the value of the labor performed and materials furnished. A suit to perfect claim of lien, or *lis pendens* in lawyerese, binds the real property for the value of the labor performed and materials furnished similar to the manner in which a mortgage or deed of trust secures the payment of a loan or note.

A claim of lien means a written statement, signed and verified by a claimant or his agent, containing all of the following:

- A statement of his demand after deducting all credits.
- The name of the project's owner or reputed owner, if known.

- A general statement of the kind of labor, services, equipment, or materials furnished.
- The name of the person, owner, or developer by whom he was employed.
- A description of the job site sufficient for identification.

A claimant is one who has made any site improvement at the request of an owner or his agent (architect, engineer, prime contractor, etc.) and has a lien upon such owner's property for work done or materials furnished. Project scheduler consultants fall into this category.

The first step in establishing a lien right is for anyone (with the exception of the prime contractor or workers working for wage) interested in establishing a lien right to file a Preliminary Notice. If an invoice from a material supplier contains the required information, it may serve as Preliminary Notice. Mechanic's liens are rights under the state's constitution and are liens on the property for "value of the labor done and materials furnished." A mechanic's lien is a legal means of recovering payment for works of improvement and site improvements by contractors, subcontractors, laborers, and suppliers of materials, including, but not limited to, new construction, construction alterations, tenant improvements, additions (works of improvement), repairs on almost any structure, or demolishing or removing improvements (site improvements). Mechanic's liens attach to both the structure and the land on which it sits.

The requirements for mechanic's liens against the project are:

- Commencement of work, however slight. There must be actual visible work, and/or delivery of materials to the job site.
- It must be shown that labor and materials became a fixture on the land, and that they are of a structural and permanent nature.
- Work must be done at the request of the owner, or his agent, contractor, subcontractor, architect, builder, etc. Knowledge by the owner of the construction is grounds enough for judgment.
- If an owner of a structure or property records a Notice of Non-Responsibility within ten days after knowledge of the construction, no mechanic's liens will apply against him unless he is the one ordering said work to be done.

Liens are filed with the county recorder's office in the county and jurisdiction in which the project is located. Liens can be filed by all original contractors, subcontractors, or suppliers who have previously filed preliminary notice. Preliminary notice is a required prerequisite by law in order to file a mechanic's lien. These liens must also be verified by the claimant (the contractor or supplier) to be true and accurate and if not correct, are considered void, therefore unenforceable by the courts. They must be filed no later than thirty days after notice of completion by subcontractors and no later than sixty days for prime contractors. If no notice of completion is filed, then all have ninety days from their finish date to file. Once filed, liens must be perfected in ninety days from the date the lien was filed. This is called a "suit to perfect lien" and it forces foreclosure of the property by court judgment of default of payment. However, the claimant is allowed by law to grant extensions of the foreclosure for up to one year.

Just having the mechanic's lien right and filing the required Preliminary Notice in a timely manner does not guarantee that the contractor will get paid. It simply means the contractor has what is termed legally a "cloud on the title," the same as a mortgage or grant deed holder. For the contractor to get its money, it must do exactly what a bank would do... foreclose. Foreclosure requires a lawsuit to perfect a particular Claim of Lien, and it must be commenced within a period of ninety days immediately after the date of filing of the particular Claim of Lien. Liens and the perfection of liens must be filed with the county recorder in the area of jurisdiction wherein the project is located. If the contractor is late on this, the contractor loses. In fact, if the contractor is late on any of these legal notice filings, the contractor loses. The courts and construction law hold a licensed contractor to a higher level of accountability and responsibility. A CPM scheduler, working for a contractor, can't stand in court and say, "I don't know, I must have overlooked my filing dates." Accordingly, ***FastPro*** allows you to make monthly planners on the network timeline master, then save them to separate files which will detail and track your document filing deadlines.

For a licensed contractor to get its money and perfect a lien, it is necessary to file a lien foreclosure action within ninety days of recording the Claim of Lien. A suit to perfect lien, which establishes a court date to sue for foreclosure on the property to settle claims of unpaid money due the contractor, is this necessary second step. If more than ninety days are needed in which to foreclose, credit should be given to the project owner. Within ninety days after his offer of extension, a contractor must foreclose or file for an extension. The contractor is the only entity that has the right to give the owner an extension for up to one year. Recorded extensions may be granted to the owner by the above procedure, but the contractor must foreclose within a year after work is completed. The action should be brought to trial within two years after commencement of suit to perfect lien.

Notice To Owner

The contract document, Notice To Owner, describes in non-technical language, in clear and easy to understand words, the State's mechanics lien laws and the rights and responsibilities of both the owner and the contractor. This document also advises the owner to require payment and performance bonds from the contractor to protect the owner from the contractor's not finishing the job and not paying his subs and suppliers. A bond is an insurance policy that has been taken out to guarantee a certain event will happen or not happen.

A performance bond is an insurance policy underwritten by an insurance company to ensure that the contractor will perform the work as contracted. If the contractor abandons the job, the insurance bond will cover the cost of another contractor's coming in and finishing the job. A payment bond likewise guarantees the owner that the contractor will pay his subs and suppliers. This document must be given by the contractor to the owner before he or she signs the contract. If it is not included or attached to the original contract, the contract is invalid. This is a legal technicality big enough to nullify claims made against the owner or project.

Stop Notice

A Stop Notice is filed by suppliers or subcontractors who have previously filed preliminary notice in a timely manner, to the holder of funds (public agency, owner, or lender). It is a legal notice to request that funds be withheld from the general contractor. Stop Notices can be sent on public or private jobs. The Stop Notice is used on public works jobs in place of a lien, because public property cannot be liened. Any activity contractor, except the original (prime) contractor, may serve a Stop Notice.

A Stop Notice is a "lien on funds." It is a legal notice to the awarding authority or the holder of funds by a subcontractor (claimant), to withhold from the prime contractor in order to satisfy claims for labor and materials furnished to the prime contractor. When a subcontractor (specialty contractor) has contact directly with anyone from the awarding authority (government agency), the subcontractor is then the original (prime) contractor and is subject to the provisions of the law relating to original contractors.

Stop notices must be verified by the owner to be true and correct. However, the claimant on a stop notice is the contractor, not the owner. To obligate the holder of funds to withhold funds, the stop notice should be bonded for one hundred twenty-five percent the amount of the claim. This is called a *bonded stop notice* and it is the heavyweight champion of contract documents. A stop notice stops the flow of money into a project, and things get critical quickly. Upon receipt of a stop notice, the lender may withhold funds from the owner or prime contractor. Upon receipt of a bonded stop notice, the lender must withhold funds from the owner or prime contractor.

Both require a lawsuit to force actual payment. Neither mechanic's liens, stop notices, nor payment bonds impair the contractor's right to court, but rather are legal aids to recovery. The stop notice is used only by subcontractors and third parties, such as suppliers. It is usually more effective than a mechanic's lien because it is used to withhold a specific amount of money from the prime contractor, which the subcontractors claims he or she is owed. Since the stop notice is used to force the lender (or holder of funds) to withhold a disputed amount of money, the subcontractor can be assured that, if any money is left in the account, it will be held until the dispute is resolved.

A stop notice is the only single means to force payment on a public works job. Public property cannot be liened, so in the event that a contractor needs to serve a stop notice on a public works job, he must send it to the awarding authority which is the project coordinator for that public works job. A bonded stop notice ensures compliance by the construction lender to withhold funds from the prime contractor to satisfy the amount claimed by the subcontractor or supplier. Without the bond, the holder of funds is not obligated to withhold the money.

Chapter

10

Construction Law

Project Schedule Labor Use

Construction historically is one of the most dangerous businesses to be in. No one will argue the statistics of injuries and fatalities that occur within this inherently dangerous occupation. Accordingly, there are many laws concerning your CPM project schedule's use of human labor that you must be aware of. During your development stage of the project's CPM network schedule, units of labor must be broken out for activities' work items' duration. Regulated use of union and nonunion labor directly affects the time necessary to complete a task or work item. In turn, that directly affects the overall activity time. Accurate time estimates for labor must reflect area adjustment for your project's local labor market. Each arena of the construction industry (residential, commercial, industrial and public works) must operate within the parameters of different labor codes and regulations.

Labor Codes

Your project's use of manpower is governed by the following regulations of the Department of Industrial Relations, whose function in the labor codes pertaining to your project schedule is to foster, promote, and develop the welfare of the wage earners of America. This department also administers the states' plans for the development and enforcement of occupational safety and health standards (Occupational Safety and Health Administration [OSHA]). The department has a subdivision called the Department of Industrial Accidents, which deals with workers' compensation. The following are paraphrases of the labor codes that relate to your project schedule:

- The Labor Commissioner and his authorized deputies have free access to all places of labor within your project. Failure to admit them to the project site for inspection, or the home office for record checking, is a misdemeanor and will stop all work on the project.
- Any employee who is discharged, threatened with discharge, demoted, suspended, or discriminated against because such employee made a truthful complaint against the employer shall be entitled to reinstatement and reimbursement for lost wages and work benefits. Violation of this law by any of the activities subcontractors is a misdemeanor.

- If any hourly employee of the project team or of the activities subcontractors discharges an employee, the wages earned and unpaid at the time of discharge are due and payable immediately.
- If any hourly employee of the project team or of the activities contractors, not having a written contract for a definite period, quits his employment, his wages become due and payable not later than 72 hours after termination.
- If any activity subcontractor willfully fails to pay any wages due to an employee who is discharged or quits, the wages of such employee continue as a penalty from the due date at the same rate until paid, but continued pay shall not continue for more than 30 days. In labor codes this is known as "pay goes on."
- If an activity subcontractor pays an employee by check and subsequently the check cannot be cashed, such wages shall continue as a penalty from the due date at the same rate until paid, but continued pay shall not continue for more than 30 days. The penalty shall not apply if the subcontractor can establish to the satisfaction of the labor commissioner or an appropriate court that the said violation was unintentional.
- All wages earned by any person working for hourly wages on the project are due and payable at least twice during each calendar month. Labor performed between the 1st and 15th days shall be paid for between the 16th and 26th day of the month. Work performed on the project between the 16th and the last day of the month shall be paid between the 1st and 10th day of the following month. However, the salaries of administrative or salaried management employees may be paid once a month.
- In a dispute over wages, the activity subcontractor shall pay all wages conceded by him to be due, leaving the employee the right to dispute the remaining amount.
- Every activity subcontractor employer shall keep posted conspicuously at its place of work a notice specifying the regular pay days and the time and place of payment.
- In the event of a labor strike, the unpaid wages earned by the striking employees shall become due and payable on the next regular payday.
- No activity subcontractor shall issue in payment of wages due any check or draft unless it is negotiable and payable in cash upon demand. No activity subcontractor shall issue scrip, coupons, or other things redeemable in merchandise as payment for wages.
- An activity subcontractor may not withhold from any employee any part of his wages unless so required by the government.
- No activity subcontractor shall charge a prospective employee for any pre-employment medical or physical exam, or charge an employee of the same.
- Every activity subcontractor shall, semi-monthly, at the time of each payment of wages, furnish each employee, either with a detached part of the check

paying the employee's wages or separately, an itemized statement in writing showing:

- Gross wages earned
- All deductions, including SDI, state taxes, FICA, and federal taxes
- Net wages earned
- Dates of work period
- Name of employee or social security number
- Name and address of employer

- Every activity subcontractor who pays wages in cash shall, semi-monthly or at the time of each payment of wages, furnish each employee an itemized statement in writing, showing all deductions, dates of work period, name of employee and Social Security number, and the name and address of the activity subcontractor.
- Whenever an activity subcontractor has agreed with any employee to make payments to a health or welfare fund, pension fund, or vacation fund, it shall be unlawful for such activity subcontractor to willfully, or with intent to defraud, fail to make the payments required by the terms of any such agreement.
- No activity subcontractor shall demand or require an applicant for employment, or any employee, to submit to or take a polygraph or lie detector test as a condition of employment or continued employment. The prohibition does not apply to an agency of the government.
- No activity subcontractor shall require any prospective employee or any employee to disclose any arrest record which did not result in conviction.
- Eight hours of labor constitutes a day's work, unless stipulated expressly by the parties to a contract. No employee shall be required to work more than eight hours per day or more than forty hours a week unless time and one-half is paid.
- Each person employed is entitled to one day's rest in seven. Violation of the law is considered a misdemeanor.
- Labor laws usually apply to those hours exceeding thirty hours a week. Employment that does not exceed thirty hours in one week, or six hours in one day, is considered part-time and is not protected by the same laws.
- No activity subcontractor shall force any employee to join or not join a labor union.
- An activity subcontractor who seeks to replace employees on strike or not working due to a lockout shall be permitted to advertise for replacements, if such advertisement plainly mentions that a labor dispute is in progress. A strike means an act of more than fifty percent of the employees to lawfully refuse to perform work.

- A jurisdictional strike means a concerted refusal to perform work for an employer arising out of a controversy between two or more labor unions, as to which of the unions has or should have the right to bargain collectively with an employer.
- The labor commissioner may issue a citation to any activity subcontractor disobeying the labor laws. The activity subcontractor has, upon receipt of the citation, ten days to request a hearing of contestment.
- A collective bargaining agreement is an agreement between management and labor to which they have collectively agreed following a period of bargaining. These agreements are enforceable in court.
- Where a collective bargaining agreement contains a successor clause, such an agreement shall be binding upon any successor activity subcontractor who purchases the contracting activity subcontractor's business until expiration of the agreement or three years, whichever is less.
- Professional strikebreaker means any person who repeatedly offers himself for hire to two or more employers during a labor dispute, for the purpose of replacing an employee involved in a strike or lockout. In most states, such strikebreakers are illegal.
- A minor under sixteen years of age is forbidden to:
    - Work on machines, power tools, saws, etc.
    - Work on scaffolds.
    - Work on excavation work.
    - Drive vehicles.
    - Work with dangerous or poisonous acids.
    - Work in tunnels or other types of excavation.
- No discrimination shall be made by any activity subcontractor in the employment of persons because of race, religion, color, national origin, ancestry, physical handicap, medical condition, marital status, or sex. Nothing shall prohibit an activity subcontractor from refusing to hire or from discharging a physically handicapped person unable to perform duties, or whose performance would endanger his health or the health or safety of others.
- It is unlawful for an activity subcontractor to refuse to hire or employ, discharge, dismiss, suspend, or demote any individual over the age of forty on the grounds of age, unless the person fails to meet the bona fide requirements for the job. This shall not limit the right of an activity subcontractor to select or refer the better-qualified person from among all applicants for the job.
- Activity subcontractors shall keep all payroll records, showing the hours worked, wages paid, etc., for at least two years.
- No activity subcontractor shall knowingly employ any alien who is not entitled to lawful residence in the U.S., if such employment would have an adverse effect on lawful resident workers.

- Each activity subcontractor must secure an employer's identification number from the Internal Revenue Service or from the Social Security Administration.
- Each activity subcontractor must ascertain each employee's correct social security account number and copy it directly from the employee's social security number card when he or she starts work.
- Every activity subcontractor who has one or more employees in his employ and pays wages in excess of $100 or more during a calendar quarter becomes an employer within the meaning of Unemployment Insurance Codes and is subject to its provisions. Everyone who becomes an employer is required to register within fifteen days with the state's Employment Development Department.

OSHA

The Occupational Safety and Health Administration is given the responsibility to make sure employers provide healthful and safe working conditions for their workers. To do this, OSHA is authorized to enforce effective standards, to assist and encourage employers to maintain safe and healthful working conditions, and to provide research, information, education, training, and enforcement in the field of occupational safety and health.

There are four divisions of OSHA:

1. **The Standards Board**. This is the legislative branch of OSHA that makes the laws (sets the standards) that pertain to safety in the workplace the workplace. They can be compared to the legislative branch of government.
2. **DOSH**. DOSH stands for the Department of Safety and Health. They enforce the standards that are set by the Standards Board. Inspectors and complaint investigators from OSHA are from the second division, DOSH. They can be thought of as the cops who enforce the laws and issue the citations for violations of safety and health laws.
3. **The Appeals Board**. The Appeals Board hears appeals to the citations issued by DOSH. They can be though of as the court for dispute resolution between companies cited and DOSH. Appeals to citations issued by DOSH must be appealed within fifteen days of issuance of that citation.
4. **The Consultation Board**. The Consultation Board issues safety posters and advice. This is an excellent resource for you, the project scheduler. These people are extremely helpful and will tell you right over the phone what your project needs to have to be within code compliance. Forewarned is forearmed.

When the owner asks what health and safety code compliance is going to cost on the project, and the other members of the project team are sitting there with blank looks because no one thought of it, let them sweat for a moment before you give up the information. Think the owner is going to look at you with respect for your professional attention to detail?

Whenever OSHA learns or has reason to believe that any place within your project is not safe or is harmful to the welfare of any employee, it may investigate the place of employment with or without notice. When OSHA receives a complaint from an employee that his place of work is not safe, it shall investigate within three days for serious violations, and within fourteen days for non-serious violations. OSHA is empowered to investigate the causes of a project site accident which is fatal to an employee or which results in serious injuries to five or more employees. To make an investigation or inspection, OSHA shall have free access to any place of employment to investigate and inspect during regular working hours, and any other reasonable times when necessary for the protection of employee's safety and health. If permission to investigate or inspect is refused, OSHA may obtain an inspection warrant.

The penalty for an activity subcontractor's demoting or firing of an employee for notifying OSHA of a safety violation is that the activity subcontractor will be forced to rehire or repromote the employee and pay all back pay. No employee shall be laid off or discharged for refusing to work in an unsafe or unhealthful atmosphere. Any employee of any project shall have the right to discuss safety violations or safety problems with an OSHA inspector. If, upon inspection or investigation, OSHA believes that an activity subcontractor has violated any health and safety order, it shall issue a citation to the activity subcontractor. The citation shall give a time limit within which the violation must be fixed. Each citation shall be prominently posted at or near each place within the project that a violation occurred. All posting shall be maintained for three days or until unsafe conditions have been fixed, whichever is longer. The activity subcontractor is given fifteen days to appeal the citation if he wishes to contest it. If the condition of any employment constitutes a serious menace to the lives or safety of persons, OSHA may apply to the superior court for an injunction restraining use or operation until such unsafe condition is corrected. This will cause cessation of production of the project.

Notices placed in conspicuous places within the project must not be removed. anyone using or operating any such operation or machinery before it is made safe, or who defaces, destroys, or removes such notice without the authority of OSHA, shall be guilty of a misdemeanor. Any authorized representative of OSHA may prohibit the use of a device, machine, or piece of equipment in the project for 24 hours on his finding that an imminent hazard exists. When, in the opinion of the regional manager, it is necessary to preserve the safety and health of workers, he may prohibit further use for not more than 72 hours. Activity subcontractors must provide opportunity for employees to observe "exposure-to-hazard" reports provided to employers identifying safety hazards. Any activity subcontractor who willfully or repeatedly violates any occupational safety or health standard may be assessed a civil penalty of not more than $10,000 for each violation. Any activity subcontractor who fails to correct a violation within the period permitted for its correction may be assessed a civil penalty of not more than $1,000 for each day during which such failure exists. The following types of construction work involve substantial risk of injury and require special permits from OSHA:

- The construction of trenches or excavation which are five feet deep into which a person is required to descend.
- The erecting of scaffolding more than three stories high. The demolition of any building or structure more than three stories high. Jack scaffolds, or lean-to scaffolds, are now illegal. Safety lines must be provided for workers working more than fifteen feet above ground level.

Regularly scheduled safety conferences are required by OSHA. The safety conference shall include a discussion of the activity subcontractor's safety program and methods intended to be used in providing safe employment on the project construction site. "Tailgate" safety meetings with employees are required at least every ten days. Activity subcontractors shall also hold monthly meetings with foremen and management, to discuss safety measures. These, as well as the tailgate meetings, must be documented and a copy must be given to the project's site supervisor, who in turn should record these on the field reports forwarded to you.

Public Works

Civil and public works projects can be small to gigantic. Logistics of operations alone on a government job can require a full-time crew. CPM scheduling comes from this arena, and works well in the multiple straight-line continuims typical in government public works projects. Two types of CPM scheduling which we have examined, PDM and Activity-On-Node, work best for public works project execution. These project contracts are overseen by an awarding authority who has been appointed by the public works agency to manage the project. The awarding authority is whom your CPM project schedule will be contracted with.

Your CPM network schedule will be affected by two factors in the public works arena: the quantitative survey takeoff and the activities durations and specifications from the architect and the prime contractor bidders. Once the bidding process is finalized, use of regulated labor in your CPM network schedule is at this time subject to the provisions of the Subletting and Subcontracting Fair Practices Act. Critical activity duration events you receive from subcontractors must therefore be compared to these specifications for accurate and realistic labor time estimates. Surprises here are not good. Your projected time estimates in each work item must be accurately balanced.

Lien Period Timetables

Liens must be filed no later than thirty days after notice of completion is filed by subcontractors, and no later than sixty days by prime contractors. If no notice of completion is filed, then all have ninety days from their finish date to file. Once the lien is filed, it must be "perfected" in ninety days from the filing. The accepted practice for the CPM project scheduler to track these timetable deadlines is to set up a separate network timeline file, with the following data table entered and interlinked with milestones on the network schedule.

Legal notice	Who	Where	When	Why
Preliminary Notice	Subcontractors Suppliers	To: Prime, owner, lender	Within 20 days of activity start	Secure lien rights
Notice To Owner	Residential contractor	With contract	Signing of contract	Secure lien rights
Stop Notice	Subcontractors Suppliers 3rd Parties	Holder of funds	30 days Notice Of Completion	Hold money from prime
Notice Of Completion	Owner or prime	County recorder	10 Days Notice of Completion	Limit lien time

Tracking the lien timetable is simply a matter of entering the minor milestones on the network timeline of your CPM schedule. The legal notices lien timetable should be treated just like activities on the network schedule, with early and late filing dates entered as early and late start events. Since there is no critical path to their sequence, they function as parallel activities in the computer and appear as events on the network schedule.

Chapter 11
Schedule Estimating

Takeoffs

Takeoffs in construction estimating are breakouts of units of the labor and materials involved in the job. Takeoffs in CPM are also breakouts of units; however, these units are of time. And in the schedule, time is measured in elapsed time from event to event. What makes the elapsed-time estimate for the CPM method work so well is the averaging of offsetting pluses and minuses over many activities and tasks. The estimate is the basis for the bid, which becomes the budget for the project. Without extensive planning and critical scheduling, lack of coordination will result in added costs, which quickly reduce the profits. So the estimate must be critically analyzed in your schedule planning.

CPM management is the best way to plot out and execute a project. It touches on all aspects of work and is a more effective means of totally viewing a project when compared to bar charts. However, if the estimate was done poorly, the estimate was made at the job logic level by estimating the labor hours needed to complete each work item, but no consideration was taken that actual field construction has practical constraints, and no real-world cushioning allowance was included. So float in the CPM schedule has to be used efficiently to bring all the labor, materials, and fabrications together at the proper time. But the original estimate should show no float. You can best control your estimates by comparing them with others and, if prices vary significantly, by verifying the numbers.

After running the first pass on the computer, it's a good idea to recheck the elapsed times of the critical and near-critical activities, to ensure that the time estimates are reasonable. The key critical activities are the twenty percent or so that most directly affect the project completion date. You should also consider any abnormal factors, such as conflicting workschedules, adverse weather conditions, shortages of material and human resources, and delivery dates that could adversely affect an otherwise normal elapsed-time estimate. We have seen in the business combat chapter that contractors are typically not allowed delay claims for unseasonable or inclement weather. Keep in mind, however, that field labor productivity reflects directly on elapsed-time estimates. As an example, the basic activities (with their average elapsed times in estimated working days) to construct a typical residential house in the Midwestern states are listed here:

Activity	Elapsed time
Clear site	10
Excavate basement	15
Rough walls and floors	45
Exterior siding	35
Pour basement	30
Roof	20
Rough-in plumbing	25
Rough-in electrical	20
Finish plumbing	20
Finish electrical	14
Interior painting	30
Drywall, taped & textured	27
Exterior fixtures	10
Exterior paint	30
Finish flooring	25
Roofing	20
Landscaping	30

Build your own sample network schedule for these activities by arranging the work activities according to their job logic requirements. Now draw the arrow diagram from each activity, showing connectivity and sequential flow. Next calculate the early and late start dates, with the project starting any time you wish. Now calculate the earliest completion time and the critical path. Finally, interdependencies and constraints must be identified between activities, with float used to balance constraints. This is the basic format in elapsed-time estimating in CPM scheduling. Granted, this is a simplistic example, but regardless of the size of the project the fundamentals remain the same. You'll just have a lot more activities.

Quantity Survey

For most civil engineering and public works projects in America, a professional quantity surveyor, engaged by the awarding authority handling the project, determines the quantities of materials required on the project. The breakout of these materials and their quantities are then published and made available to all bidders on the project. Each general contractor interested in bidding on that project then has the estimator work up a price list for all materials.

In bidding on quantity survey projects, all contractors are bidding on the same quantities of materials, and therefore the contractor's estimator builds a job estimate based on unit prices of those materials. The estimator then computes labor costs to install, and breaks those costs down into per-hour units. The same happens with equipment costs to install materials. In this method of competitive bidding, all contractors will be bidding on the same quantities of materials and need to keep the costs of purchase, delivery, and installation of those materials as low as possible to get the contract.

The quantity survey method of bidding is typical of larger public works projects, and you undoubtedly will encounter projects of this type in your career. To develop a project schedule for quantity survey projects, the scheduler must know the actual number of units required for each activity completion, and must check those units against the original number of units on the quantity survey takeoff. For example, if the original quantity survey called for 1487 linear feet of concrete curbing, and 1515 linear feet were actually installed, then the contractor would be paid for the additional 28 linear feet. If 1400 linear feet were installed, then the awarding authority would pay only for the 1400 linear feet actually installed and not the 1487 linear feet. originally called for in the quantity survey. This is a practical method of adjustment between what was originally proposed and what was actually installed. Adjustments are made in case of errors in this manner: Large errors require the unit prices to be renegotiated, and small errors are usually adjusted at the same unit rate the contractor originally bid.

Long-Lead Items

Long-lead items are those items or materials within activities that are going to have to be manufactured, custom fabricated - or will take a long delivery time, such as out-of-area or out-of-state transport. The best time for subcontractors' ordering of long-lead items is after review of the plans and specifications. Many contractors will wait until the last moment to order items of these types because they usually require partial down payment, and no one wants to have out-of-pocket expenses until necessary. However, if you allow subcontractors who have long-lead items in their activities to order those items only as needed, the subcontractor is running the risk of down time if the items are then unavailable or on back order. Making sure long-lead items are ordered after going over the plans and specifications ensures timely delivery with no activity delay further setting back other critical path activities. We ensure that subcontractors have indeed done this by requiring that copies of their purchase orders to suppliers of long-lead items be given to the project manager within a specified time after the signing of contract. This requirement should appear in the contract specifications, and guarantees that items are ordered in advance.

Cost Tracking

One of the many crucial tasks of the project scheduler during project operations is that of monitoring the project field service costs, as separate and distinct from the architect's design costs. The amount of the field costs is a function of the size and classifications of the on-site field labor forces assigned to the project, field office overhead costs, materials and supplies, support services from the home office, vehicle leases and fuel charges, the field office's share of corporate general and administrative (G&A) costs, outside consultants or contract services, and job profit.

By tabulating the monthly accumulations of budget and actual field service costs and then plotting each amount on a time-versus-cost chart, similar in form to the traditional S-curve chart, you can make a visual comparison that clearly shows not only the status of the contract at any given time, but will also show the project manager or owner any change in trend toward either a savings or a cost overrun. By also plotting a curve representing the amounts invoiced to the owner for such field services, you can provide an additional dimension of usable data to the owner.

To prepare and maintain the cost-tracking chart on the computer program, enter the items into the Cost By Activity Number program file. This sort will calculate the estimated, current, and projected due costs for every item you enter, in every activity you're using. Accurate data for your cost tracking requires regular inputs from the site supervisor on all field costs, and hours of work in each classification at the project site. The project scheduler arranges for all such field data to be received from the project manager at the end of each week. Typically, the monthly pay estimates of the project's contractors' work for progress payments are submitted by the 25th of the month, and all submittals are to be in the office for payment before the end of the month. Billings from the architect to the owner, however, usually are based on the closing date at the end of the month. Therefore, weekly tabulation of these data should not interfere with the project manager's review of the contractors' pay requests. The cost tracking should be done on a unit basis in all activity items. These are then used to produce the audit trail.

One of the major advantages of computerized CPM is the ability to link items' and activities' cost to the network schedule milestones of payment. Most major retail computer programs link cost tracking linked with inventory control (the barcode on the products at the grocery read by laser scan and entered into the computer for readout). For example, as the item is sold to you in the grocery store, it is also subtracted in the inventory. Milestones exist within such programs to signal the purchasing program to buy more of those items when inventory levels are low. All such information provides an audit trail of each item. By setting up the Cost By Activity Number in your software program, you create the schedule's audit trail.

Audit Trail

Figure 11-1 shows the Cost By Activity Number sort. The factors used by the computer to build a cost-tracking database used in computations of the audit trail are:

- The activity's assigned *i-j* number
- The activity description
- Start event of the activity
- Finish event of the activity
- Line float of the activity
- Available (committable) float
- Negative float (if any)
- Prior event
- Next event
- Activity cost expense data
  - a) Estimated
  - b) Current
  - c) Projected Due

The computer uses the float data entry to provide calculations under the "What-If" pulldown menu. This forecasting advantage allows the project scheduler to input other numbers into the cost and time columns for computation. The next step is to critically analyze what those changes would bring if initiated immediately. The subtotals of each column can either be totaled or averaged, or both. Printouts can be run at any time, to freeze a particular cost scenario that appeals to the owner.

After critical analysis of all options is complete, the program returns to the original numbers without disturbing any other data. This is especially useful in construction-loan, interest payment scenarios. These financial wraparounds allow the owner to maximize the potential of the loan funds without dipping into personal lines of credit, thus keeping working capital on hand and available. This overall process is known as *trends analysis*. By entering the relevant data on the Cost By Activity Number sort, the computer now links the activities network with the costs and expenses necessary to run the network. In addition, owners using CPM systems have the advantage of control over those costs before they are paid, by having the project scheduler run trends analysis forecasting. The critical art of monitoring the project field service costs now linked to the project schedule allows you to separate those cost expenses from the architect's design costs. The Cost By Activity Number sort tracks and calculates the classifications and the amounts of the on-site field labor forces assigned to the project, as well as the field office overhead costs. An entire database can be built to show costs of materials and supplies, support services from the home office, vehicle leases and fuel charges, field office share of corporate general and administrative (G&A) costs, outside consultants, professional legal services, professional engineering or contract services, and separate or combined cost-to-profit ratios.

Audit trail trends analysis is an important function of schedule estimating that will provide your client with one of the prime advantages of CPM, that of cost control and the ability to foresee the commitment of resources before they are needed. Figure 11-1 shows a typical software program layout for an audit trail.

FastPro CPM Scheduler

Your Company Name Here

Prepared For *Your Client's Name Here*
Project: *Project's Name Here*
File Name: *Your Computer File Name Here*
Description: *Project Description Here*
Our Invoice: *Your Company's Job Number*
Client Invoice *Owner's Job Invoice Number*

Audit Trail Tracking

Page: 1 of

Data Date:
Run Date:

Date Linked	Item Number	Item Description	To:	From:	I-J Number	Date Appv'd	Date Req'd	Date Sent	Check Number	Order Status	Description Value

Chapter

12

Contract Conditions

General Conditions

"Boilerplate" is a nautical term that refers to the strong, often thick material (usually steel) that forms a pressure vessel, such as a boiler. In construction, the general conditions are like "boilerplate" that provide containment precedence for dispute resolution among the parties of a project, by anticipating most of the areas of discussion or dispute that might arise and providing for an orderly method of resolving each case.

In contracts, the general conditions serve to "boilerplate" the vested interests of the owner and lender, by providing prearranged avenues of precedence in order to keep the production in motion during times of dispute resolution. Every liability, possibility, and probability must be identified, critically analyzed, and prepared for. Contingency plans, insurance policies and bonds should be obtained as collateral for every dollar of your company's investment in the project. Every clause in every one of your contract documents should be worded, by a contract-specialist lawyer, so that its contents focus squarely on protecting the production and timely completion of your CPM project schedule.

In addition to the working relationship provisions they contain, the contract's general conditions are not limited to the production portions alone. Typically, all things that are bound into the contract's general conditions are considered part of the primary contract document. These may include the bid bonds, agreement contracts, performance and payment bonds, and conditions of the contract. By law, construction contracts encompass general conditions, plans, relative contract documents, specifications, and technical specifications addenda.

Generally, construction contracts can be broken down into three categories, though they are not necessarily arranged in the same order on each job. These categories are:

Part I, Bidding and Contractual Documents

- Notice of Invitation to Bid
- Instructions to Bidders
- Proposal Bid Forms

- Bid Sheets
- Contractor Certificates (Licensing & Surety Bonds)
- List of Subcontractors
- Bid Bonds
- Non-Collusion Forms
- Agreement Contract
- Performance Bonds
- Payment Bonds

Part II, Conditions of the Contract:
General Conditions
Supplementary General Conditions

Part III, Specifications and Specifications Addenda
(Architect or design professional herein provides technical sections covering the various parts of the project.)

Each of these items has further details, but an in-depth examination of each is beyond the scope of the project scheduler's responsibility and falls instead in the lap of the owner and project team. The architect or design professional is responsible for supplying technical specifications as well as the design plans. These contract specifications and their addenda, are added to the contract's general conditions to explain in detail the various responsibilities of the architect or design professional regarding the review and approval of shop drawings, as well as the parameters of design to production liabilities. Project schedulers need to be aware of all this information to prevent unclear or incomplete details specifications from causing delays in their schedule.

For schedulers, reviewing the general conditions, specifications and specifications addenda is a critical step to ensure contingent activity workarounds, and that long-lead items do not burn up valuable float time, thereby placing your completion date in jeopardy. Prime contractors also need to function within the general conditions responsibility as they subcontract to the related activities subcontractors for each contract specification, per plans and specifications

Contract Specifications

Contract specifications are not limited to the technical portions alone. Typically, everything that is contained within the specifications document is referred to as "the specifications." These may include the notice of invitation to bid, the bidding documents, bid bonds, agreement contracts, performance and payment bonds, non-collusion affidavits (where applicable), conditions of the contract, technical provisions of the contract, and provisions of the contract CPM network schedule. If you encounter plans calling for a book of standard specifications by reference, this may signal a specifications writer who was not proficient at detailing specifications. Use caution here. Every project manager, scheduler, contractor, and inspector who has spent a few years in the trades has encountered specifications that were questionable at best. (I'm being generous here because this is an understatement.) Construction management professionals will tell you that not all specifications writers are knowledgeable in all the subjects they are writing specifications for. Sometimes they're just flat wrong.

If an error in the specifications is not caught by the responsible subcontractor before the work is done, your schedule will be pinched by the dispute resolution necessary to unscrew the situation. It happens, and is one of the things you need to be aware of that can stall an otherwise viable schedule. A comparison of the Finish Schedule with the Specification Index will expose potential duplications and/or omissions. It will indicate whether the design process was completed in an organized and sequential manner, or put together by different people who never worked together. Any discrepancies discovered in this review will be fairly obvious.

Comparison of the Specifications to the Finish Schedule includes analysis of the schedule's headings against the categories included in the technical specifications. You're looking to confirm that each item is noted and accommodated in the technical specifications, and that each item is included only once. Should any discrepancy or duplication be discovered, the architect or design professional must be immediately notified of the error. Some design professionals have tried to limit the extent of their liabilities by including exculpatory clauses in the specifications addenda, which usually read something like this:

"Plans and specifications provided are complimentary. The contractor is responsible to provide all work shown, whether or not adequately described in these specifications, and all work described in the specifications, whether or not specifically indicated on the plans, to code compliance and current accepted workmanship standards." If you see something like this in the specifications addenda, cringe. You're looking at what project managers call a "weasel clause." The design professional was not an expert in that technical specification and/or didn't research it thoroughly enough. And if it wasn't worked out in the design process, the only other place it can be worked out is in production, blowing your project schedule and costing lots of money. Unless, of course, you're smart enough to be looking for these problems ahead of time. I equate weasel clauses to dancing. Every time I see one in the specifications addenda, I know it's time to go cha-cha with the architect.

Specifications and their addenda are needed in the contract documents to provide comprehensive technical explanations of project systems and their function. They must be complete for the production process to flow smoothly. If there is an absence of clear definition and delineation of these responsibilities, claims and lawsuits are the outcome. As a member of the project team, the project scheduler needs to be aware that the courts have held that the basic design responsibilities listed below are considered the architect's contractual obligation, whether expressed or implied, within the specifications or specifications addenda:

- Code compliance
- Technical accuracy of all documents
- Producing all plans and specifications
- Specific design (not to be substituted for design criteria)
- Interpretation of the documents
- Standard workability of design, as per specs
- Shop drawings submittal review and approval
- Prompt and timely response in review approvals as to not delay work
- Evaluation of the work
- Due diligence in judgment and professional expertise

In addition, the courts have held that the first responsibility of the design professional is to indicate clearly and completely all work in sufficient detail, on the plans and specifications, as to adequately describe the parameters of each of the specifications. So this ultimate responsibility to comprehensively and adequately describe each construction specification lies with the design professional who generates the plans and specifications, regardless of any exculpatory clauses. Disallowing such clauses, the courts have held that design professionals are ultimately responsible for technical specifications and their related responsibility identification. Requests For Information (RFI) are sent from the job site back to the designer for anything overlooked or incomplete in the specifications, by the field contractors. When too many RFIs are sent back to the architect that signals any experienced developer that it's time to hire another architect on the next project.

Specifications are not limited to the technical portions alone. Typically, everything that is contained in the specifications document is referred to as the specifications. These may include the notice of invitation to bid, the bidding documents, bid bonds, agreement contracts, performance and payment bonds, non-collusion affidavits (where applicable), conditions of the contract, and technical provisions of the contract. Contract documents, by law, encompass plans, relative documents, and specifications.

Shop Drawings

Shop drawings are detailed blueprint and schematic drawings that describe how a subcontractor plans to install and/or manufacture the project's equipment components that relate to their specific activity. They are prepared by the activities' subcontractors and suppliers, and are used in the contract to show how their installed equipment and work will meet specifications established by the structural engineers and the architect.

Under the most common arrangement in current contract conditions, the prime contractor (or project manager) and the architect share the responsibility for shop drawings review. The shop drawings reviewed by the prime contractor reveal if the equipment to be installed and work to be done will meet the design criteria. The prime contractor then stamps them with its approval, and sends them for processing to the structural engineers and architect. These design professionals will then also review and approve the shop drawings and return them to the prime contractor. In doing so, the design professionals authorize the work to proceed in the manner detailed in the submitted shop drawings, or corrections needed to be resubmitted.

It is at this point of contractor-designer reviews that areas of responsibility must be made clear to all parties. If these contractor and designer responsibilities are not spelled out in the contract conditions, the lack of responsibility will delay the shop drawing approval process. Each party will be free to decide for itself how much responsibility it wishes to absorb. Obviously, the answer is as little as possible. Often, the subcontractors include less information than the designers require. The design professionals then return the submitted shop drawings to the prime contractor for correction and withhold authorization to proceed pending additional shop drawings.

The same condition exists with the design professionals, in that, they also wish to limit their professional liability and attempt to narrow the scope of their shop drawings reviews in the contract. The combination of both parties shirking responsibility contributes to substandard shop drawings. This leads to increased confusion among all parties, resulting in exactly the opposite effect, the assumption of even more liability. The court has held that it cannot be argued that the contractual roles of the structural engineers or the architect are in an evolutionary phase created by changing contract structures. The legal precedence here is that the design professionals' responses can compromise the thoroughness and attention to detail necessary for public safety in using the construction. Prime contractors often try to place much of the burden on the design professionals.This, in turn, invites confusion and substandard construction, ultimately leading to dissatisfaction at every step of the process. Historically, this effect increase liabilities for both the design professionals and the prime contractor.

The architect is normally responsible for approving shop drawings and reviewing them for their conformance with the original design concept. The difficulty lies in the fact that there seems to be no clear or consistent description of precisely what the designer's responsibilities in the shop drawing review and approval process include. To make matters worse, some design firms treat the shop drawing approval responsibility with less importance than it truly deserves. While it may be true that many designers do understand the critical nature of information incorporated in shop drawings, and treat them with respect and proper attention, some do not. Time and expense pressures in some design offices create a temptation to assign the shop drawing review task to a junior member of the staff, with lesser experience in such matters.

In this case, the likelihood of errors in detecting deviations from conformance with the original design concept increases. This lack of clear definition and assumption of approval responsibilities has left many design offices, and their respective liability insurance companies, practically determining for themselves what their responsibilities will include. In addition, many architects and engineers have attempted to limit their liabilities by avoiding the use of the word "approved" in their shop drawing remarks altogether. Contract conditions phrasing like "No Exceptions Taken," "Furnish as Submitted," "Examined," or the real cute one making the rounds currently, "Not Rejected," have now become the rule rather than the exception. Moreover, the shop drawing stamps have been supplemented with elaborate language explaining what is and is not being done, in an effort to define the review process and minimize the design professionals' liability.

There is legal opinion, however, that supports the idea that the designer's stamps may actually increase their liability, if their contract obligations for shop drawing review exceed the limits of the language included on the stamp. The owner and contractors rely on the designer's approval responsibilities as defined in the contract. If a designer operates in a narrower view as defined in the loose language on a stamp, it is a clear admission that the designer is doing less than he or she is contractually obligated to do. The courts have upheld that expostfacto language in review stamps do not relieve public safety concerns in contractual obligations.

Beyond the basic designer/contractor approval responsibility clarification, another project scheduler's concern regarding shop drawings is designer response time. Common language in construction contracts notes that a designer will "review and approve shop drawings with reasonable promptness so as not to cause a delay in the work." "Reasonable response" is another term that lacks precision. "Reasonable" becomes defined by trade practice in the project's geographic location. If nothing better is available in the existing contract language, establish the definition of "reasonable time" at the very first job meeting. Ten working days (except in unusual or adverse conditions) is standard. It's always best to pin it down early. To emphasize the precise requirement, indicate on each submission transmittal the exact response date that approval of the respective item is required by before the project is impacted. A review of the specifications may establish the procedures for shop drawing submissions. Is there a special stamp required with the project name, and other identifying data that must be used on each shop drawing submission? If so, the stamps will have to be procured. How many copies of shop drawings are required for submission? How many copies of product data or equipment catalog sheets are required for submission? How are samples to be handled? Are all the types of shop drawings to be sent to the architect, or will structural drawings be sent directly to the engineer with an information copy only to the architect? All these procedures should be clarified in the contract general conditions.

Long-Lead Items Purchase Orders

Attention must be given to the longest delivery items (long-lead items) on the schedule to see which ones fall on the critical path. If there are several large field-erected items with varying completion dates, you might even break the most important item into individual units to track the latest one. The same critical analysis applies to time-critical subcontracts.

Long-lead items need purchase orders confirmed in advance by all associated subcontractors. Purchase order copies should be supplied by subcontractors and suppliers, in advance, to the project scheduler, to maintain adequate lead time between ordering and confirming the order in the CPM network schedule along with delivery within critical event windows in that activity. When shop drawings are sent to the architect or engineer, a note should be entered on the accompanying transmittals, indicating whether any are to be expedited or if there is an order of importance. The project manager must monitor the shop drawing log from time to time, to make sure that the drawings are being cycled in a timely manner.

The shop drawing log should show what action has been taken on each drawing when the drawing is returned. Although the transmittals forwarding the copies and the transmittals accompanying the returned drawings will be the documents of record, the shop drawing log will highlight the activity. The log should contain a column showing the disposition of the drawings after they have been reviewed by the designers. If any drawings are disapproved, they would probably be returned to their originator. If approved, they will be distributed to their originator, with copies to any concerned subcontractors, a copy to the field, and one for the office file.

Chapter

13

Preconstruction Planning

Progress Schedule Regimentation

At this stage of the preconstruction project schedule planning your agenda should be to compile a complete list showing all milestone events and their respective dates. These should include all meetings, prime and subcontractors progress payment schedule, progress inspections and testing requirements, equipment delivery dates, long-lead items delivery dates, submittals, and project phases completion dates.

As contracts are awarded, all the subcontractors should be asked to submit their own activity schedules to see how they fit into the overall CPM schedule, and to submit a list of proposed shop drawing submittals and delivery dates of the equipment after receipt of approved drawings. When these schedules are received, the project scheduler must review them to ascertain that the equipment delivery dates fit into the overall timeframe of the job. If they do, then they should be incorporated into the bar chart schedule and distributed to all subcontractors. Subcontractors should be given the complete project schedule containing their portions of work, and they should be asked to comment on the length of time and sequences allotted to them. With their input initially, and regular updated status reports, the schedule will be more accurate.

Once the scheduler has identified the activities work items at the subcontractor microcosom level, the activities listed on the primary report file (Sort By Activities) are then broken out from the Sort By Activities timeline and transposed onto the accelerated S-chart timeline. This, in effect, shifts the production network into double-time. What happens at this transposition point of superimposing the Network Timeline S-chart on the Sort by Activities bar chart is the point in your development of the schedule where this opens up details subtasks and work items within the activities and thus, windows of float. The duration events opening and closing begin to build a graphic fabric that should reflect the job logic. Bottlenecks appear and workarounds are developed. The overlaid events and milestones become workable and the job logic of the network schedule flows in an organized manner.

The most common cause of a schedule's failure to meet its cost budget is design-construction contracting in the design-lump-sum bid contract. Typically, the owner turns over too much decision-making authority by not separating the design phase from the quantity estimated takeoffs. The prime contractor at this point is not only the authority in the field work but is also in charge at the administrative level. Without shared control over all aspects of production, the natural tendency is for the contracting parties to fall into an adversarial rather than a cooperative mode. Such a scenario harms all participants, especially the owner. Owners must make an overall project schedule to check on the viability of keeping the strategic end date. They then must keep track of the overall project float to see that each major project contributor uses only that float which has been allotted to that activity. If float is shifted for any reason, the total project float must be evaluated in the new context.

During the precon, the project scheduler must ensure that all work contracts specify that ordering materials shall be done after contractors review plans and specifications. Visiting a job site is not sufficient to see all of the details of the job. If materials or items are allowed to be ordered on an "as needed" basis, delays will occur if the materials or items are unavailable or on back order. Your project schedule, no matter how well done, will be handicapped if this important step is not taken at the precon.

Bid Award Prior to Commencement

Certain things must occur once the best bidder has been determined to protect the optimistic environment for the project's production schedule. Prior to awarding the contract, the project team holds a preconstruction conference with all contractors. At this meeting, they:

1. Closely evaluate the job logic of subcontractors' schedules when they are presented for approval. Check their float numbers against the master schedule. Look for numbers in the timescale that don't add up.
2. Make subs prove that any overly optimistic projections are realistic within the master schedule.
3. Don't allow any prime or subcontractors to use up their allotted float early in the master schedule. Cushions and margins for error should be held back until the finish event, and preferably, not used at all.

4. Monitor their schedule reports carefully; check each report for accuracy, move quickly on any discrepancy.
5. Establish a precedence for checking actual progress in the field.

The project scheduler prepares an operational production schedule that the project manager uses for the weekly scheduling meetings held in the field office, so the operational schedule must be detailed in activity and subactivity reports to serve that objective. Operational schedules have a more detailed work breakdown structure than an owner's schedule, and are better suited for accurate field reports.

Once the contract is awarded, the project begins. Now everyone involved will be playing beat the clock with your schedule. Buckle up and be ready to ride tall in the saddle when it comes time to ramrod your schedule. From the moment the chute opens until the pay bell rings at the end, it'll be a busy rodeo.

Progress Schedule Regimentation

Progress schedule regimentation serves to set the precedence for all milestone events and their relative events and durations falling within the project's schedule. As stated earlier, and repeated here for ease of reference, these are arranged in chronological order to visually reflect the project's job logic. This type of sequential task completion listing is known as progress schedule regimentation. Progress schedule regimentation should include:

- All meetings, both home office and field
- Progress inspections, both municipal and site supervisor
- Engineering and testing requirements
- All long-lead items delivery dates and purchase orders from subs
- Equipment delivery dates
- All submittals due dates
- Prime and subcontractors progress payments schedule
- Project phases completion dates

Shop Drawings Log

A shop drawing log should be started so that when the shop drawings start arriving from subcontractors and suppliers, each drawing can be properly logged into the computer for tracking. Not only will a log of shop drawing activity allow a project scheduler to keep track of what drawings have been received, it will also show where the drawings have been sent and how long they have been there. The project scheduler needs to prod each subcontractor and supplier to submit their drawings promptly. At the first job meeting, the major subcontractors and/or material suppliers should be requested to submit a preliminary shop drawing submission schedule.

The log should include the major pieces of equipment for which shop drawings are required, and the anticipated date when each drawing will be submitted. The subcontractor and/or supplier should also include the approximate delivery date of equipment after the approved shop drawings have been returned to them. Once shop drawings have been received, the next hurdle is getting them approved in a timely manner. The project manager and the owner should review the incoming shop drawings before the project scheduler logs them into the computer, to determine whether they conform to the project specifications and requirements. If compliance is questionable, contact the party who made the submission to re-examine the shop drawings to verify contract specifications. If there are deviations, it might be best to note them before submitting the drawings to the architect. It is at this point that the project manager must establish credibility with the architect and engineers and must show that the shop drawings are being reviewed for compliance with the plans and specifications and are not merely being passed through without any scrutiny whatsoever.

Transmittals

A typical shop drawing is used on all long-lead and fabricated items. In addition, a separate file should be kept for each trade that will be submitting large numbers of drawings, such as structural steel, plumbing, HVAC, sprinkler system, and electrical work.

There are several software programs on the market today that allow the project scheduler to create a fax modem transmittal forwarding the shop drawing to the architect or engineer and transferring this information automatically onto a shop drawing log. The information inserted at the time of preparation is stored and transferred to the shop drawing log. Although the shop drawing log contains material from various subcontractors, separate computer files and summary report sorts can be created for individual trades.

Care must be taken to discern which subcontractors should receive informational copies of shop drawings. For instance, when a mechanical subcontractor is being sent an approved copy of its boiler shop drawing, the electrical contractor should have an informational copy in order to confirm the line voltage requirements. All too often a piece of equipment is ordered with electrical characteristics at variance with the voltage requirements shown in the drawings. If an error such as this can caught in the shop drawing stage, there may be little or no additional cost involved to make the equipment compatible with the building's electrical system.

Schedule Planning

Everyone has their own style of doing things. If you've been doing this awhile, you have your own ways and procedures that work for you. If not, this is the basic way of going about planning your network schedule:

1. Create milestones in the timescale. Planning here establishes expectations for activity start and finish events and float for your project schedule. It requires project-wide conception and project team commitment to milestone events.
2. Establish CPM priorities. CPM planning requires you to consider primary activities critical to the project, policies for ensuring those activities will be done within that window, contingency plans if they aren't and resources dedicated to unscrewing the situation, and information about the activity's relative importance to the starting event of the succeeding activity.
3. Predict problems. Schedule planning should predict problems and production bottlenecks in activities and operations. List them out. This invites the creation of controls and evaluation procedures.
4. Solve those problems. Planning here allows you to evaluate what doesn't work and to consider what might be done to make it work. With a plan detailing your assumptions and expectations, it is much easier to identify what went wrong when problems arise in the project schedule. Even guessing here will allow you to check and improve areas of concern within the schedule. This impresses owners immensely.
5. Profitability. Schedule planning needs to look at cost and cost control to force consideration of the activities' costs and the owner's cost before committing funding resources. Here the best plan tends to be progress payments linkage to phase milestones. It imposes a payment upon successful completion of each activity on time and on budget contingency upon each activity subcontractor. This also allows for the inspection of workarounds.

6. Motivation. Planning also encourages full participation from the project team and motivates everyone concerned, including you, to put forth their best efforts to a positive outcome in the project schedule.
7. Cost control. Planning controls costs by creating a specific type of report sorts for each project team member. The project manager and site supervisor need periodic reporting, the bank or lender needs audit tracking, the owner wants similar comparisons and cost-effectiveness evaluations. This is a big area of concern to all owners and is one of the major advantages to the owner of having a professional project scheduler on the project team, because architects and design professionals traditionally have shown casual attitudes toward cost control for owners.
8. Schedule development. Planning is the only way to put milestones into the operational phases of the schedule accurately. It requires considered decisions from the owner and all members of the project team, rather than planning through default. Schedule planning should focus on balancing activities completion needs, demands for resources, and milestones timescaling.

Chapter

14

Project Operations

Recycling the Schedule

Quarterly, the CPM schedule should be recycled. This means to put it to critical analysis and cook it. Float, elapsed-time estimates, summary sorts, and critical paths and trends analysis forecasting take first priority on the agenda. Critical path activities that met scheduled events need to be analyzed for the characteristics of success that can be incorporated into future project schedules. Critical path activities running in negative float need to be straightened out.

A computerized CPM system simplifies recycling the schedule. Besides the traditional quarterly reworking of the schedule, recycling becomes necessary whenever schedule deviations accumulate to a point at which some of the intermediate milestones are in jeopardy. Recycling involves revising any target dates that may have slipped beyond repair, perhaps because a significant change in scope has occurred. Exercising some what-if options allows the project scheduler to make the best revisions.

Having activities with negative float is a sign that the schedule no longer is achievable in its present form. Recycling the schedule should not be confused with the monthly progress evaluation. Recycling is necessary at the end of the quarter if your short-term milestones are not being met. The recycling procedure critically analyzes how to get the project back on track without extending the completion date.

When recycling the schedule, keep juggling and rechecking the resulting logic diagram for the best job logic. Because network scheduling, especially one in crisis, is in motion and the variables are constantly changing, professional project schedulers are paid $300 an hour and up, plus all expenses, for saving a sinking project. Jumping into the fire to rescue a runaway CPM schedule pays well, but it is much akin to jumping into a mixer-blender and having someone hit the puree button.

But one must start somewhere, and I have found from experience that the best place to start recycling is with the network's job logic. Remember, the logic diagram is the central command center of the CPM schedule. If the program logic doesn't wash there, that's the beginning of the end. Nothing else, from the logic diagram outwards in the network will be running in sync. Fix this crucial component of the schedule first. Then move into the next stage of schedule recycling, which is operations analysis.

The simplest scheduling recycling is concerned with *time* use. Thus, you might allocate ten hours each week in project cost-tracking and audit trail, and twenty hours to sorts and their communication with appropriate project team members. You might dedicate the last day of the month to end-of-month sorts and the summary reports, and an hour every Monday to pre-week schedule management planning.

A more complex scheduling recycling system compartmentalizes by *activity function*. Calculating the percentage of each activity's progress towards completion is the responsibility of that activity's subcontractor, but the summary report of that progress is the scheduler's responsibility. These more complex systems tend to be (or become) hierarchical. They simplify delegation of each activity's responsibility and according sorts but de-emphasize total project objectives and are slow to adapt to changes during schedule recycling. In planning the schedule activities and operations analysis, it is important to reflect the objectives of the owner and the project team. The degree of project team control versus prime contractor and subcontractor empowerment, and the overall project matrix network, will determine in significant part whether you will reach the project's phases milestones on time. This is a trade-off or averaging tool the scheduler uses between timescaled activity events and activity control, to accelerate or constrain the activity.

Cost Monitoring

Separate and distinct from the architect's design costs, the project scheduler must monitor project field service costs during the project. Field costs are a function of the size and classifications of the on-site field labor forces assigned to the project, field office overhead costs, materials and supplies, support services from the home office, vehicle leases and fuel charges, the field office's share of corporate general and administrative (G&A) costs, outside consultants or contract services, and job profit.

By tabulating the monthly accumulations of budget and actual field service costs and then plotting each amount on a time-versus-cost chart, similar in form to the traditional S-curve chart, you can make a visual comparison that clearly indicates not only the status of the contract at any given time, but will also show the project manager or owner any change in trend toward either a savings or a cost overrun. By also plotting a curve representing the amounts invoiced to the owner for such field services, you can provide an additional dimension of usable data is provided to the owner.

To help prepare and maintain the chart, regular inputs are required from the site supervisor, on all field costs and hours of work in each classification at the project site. Arrangements should be made by the project scheduler to ensure that all such field data is received from the project manager on a regular, scheduled basis at the end of each week. Typically, the monthly pay estimates of the project's contractors' work, for progress payments, are submitted by the 25th of the month, and all submittals are in office for payment before the end of the month. Billings from the architect to the owner, however, usually are based on the closing date of the end of the month. Therefore, tabulation of these data should not interfere with the project manager's review of the contractors' pay requests.

Schedule Operations Analysis

Interlinked operations of activities and timescale systems are the basic structures of your CPM network schedule. They include everything from materials delivery scheduling to the entire scheduling process itself. The operations analysis is unique to each schedule and each project. You should review each element of your schedule in planning but shouldn't, in the interests of time management, revise or develop a full plan for each activity and interlinked operations in every planning cycle. Look for the critical path activities and interlinked operations that are most in need of attention or offer the greatest opportunity for development of shorter events through fast-tracking.

All schedules have the following operating systems:

- **Planning**. This is the process which you initially begin when you first start building the project schedule.

- **Organizing**. In complex projects this is reflected in network scheduling diagrams, work flow charts, and sorts.

- **Controlling**. Controlling is the process of creating expectations and evaluating progress effectiveness. This involves the time analyzing of sorts and making appropriate changes in the schedule's recycling phase.

Another important area of Schedule Operations Analysis is the determination of your network schedule's capacity and efficiency. To make cost and performance evaluations it is necessary to know what your project schedule's current performance limits are. Capacity study is based upon an understanding of the tasks that are devoted to the activity. Good CPM schedulers continuously monitor and improve their capacity assessments. The initial assessment can be done through hypothesis or best "guess-estimate," but it is far more accurate if based on the company's historical data from similar projects. As an example, consider this simple illustration. For a single subcontractor providing services for one activity, the elements of capacity determination are:

- The number of hours in a desirable work week... 40
- The average number of hours of reporting effort to produce field reports on that activity... .5
- The hours per week of administrative time for the scheduler to produce the summary sort for that activity... 4

Weekly productive capacity for that activity is 36 hours (40 less administrative hours) divided by 1.5 (1 hour of computer processing plus .5 hour of field reporting).

Net capacity = 24 hours per week. This is the basis of the schedule's capacity and efficiency. The trick to matrix networking productivity improvement is to follow up with the following questions. Answering yes to any of these questions brings more CPM to the project.

1. How can I check the accuracy of my timeframe assumptions? Perhaps a daily time allocation study?
2. Can I improve my matrix network efficiency?
3. Can I reduce my schedule's administrative hours in controlling?

Daily Field Reports

Daily field reports are the fundamental documents that record actual job progress, together with all conditions that affect the work. They provide the progress reporting basis for the actual results of the schedule as compared to projected progress. They begin an organization's standardization of reporting at the source of causes and effects -- the point of production. If actually prepared daily, they are generally considered to be the best sources of job information. This is because they are supposed to be prepared immediately as the information is being generated, with no appreciable time lapses. The inclusions and descriptions are fresh in everyone's mind. This facilitates complete details of the work. In addition, they are prepared by those with authority and responsibility for the work. They have an interest in the accuracy of the information and have a significant incentive to maintain the report's accuracy and completeness. These are the people who actually witnessed the work and are recognized to be the most qualified to describe the reported facts. Because they will be the most detailed, accurate, and complete records of all job site events, the daily field reports will be used to resolve any dispute about what actually happened on the job. They are so valuable because of the wealth of information recorded in summary form. This data includes:

- Work and activity descriptions, separated as to the physical location and extent
- Labor force, broken down by subcontractor, locations, and major activities
- Equipment used and stored by the various activity subcontractors
- Site administrative staff and facilities
- Weather conditions and temperatures at key times of each work day
- Change order work accomplished, with relevant details
- Photographs taken during the day
- Visitors to the jobsite
- Meetings, discussions, commitments, and conversations

You will note that the daily field report in ***FastPro*** is two pages in length. Never use the back of a field report to continue information, because it will be overlooked when faxing, during a review, or photocopying. Because it is impossible to tell in advance what will become the most important information, all categories must be accurately maintained. It is up to the central office project administration, whatever the exact authority structure within your organization, to regularly police the reports. They must be checked often enough to ensure that all required information is being properly recorded.

What is more, the requirement that the reports be completed *daily* cannot be stressed enough. If they're allowed to be lumped together for completion at the end of the week, the accuracy of the information quickly degenerates. The completeness of the data also is dramatically reduced. When on site, you should visibly refer to the reports often, during conversations, issue research, and so on. Let your field personnel know that the information is depended on and used, and that their input is extremely important. Even if you don't really need to at the moment, come up with an excuse for looking at the reports. The minute that the field personnel begin to get the feeling that no one's looking at the information, the data will become very thin indeed. Have the reports printed on two-page carbon or NCR paper. The original is to be sent to the central office, with the copy retained in the field office. Originals should be faxed over at regular daily times, with the original sent or picked up for recordation. Requiring daily deliveries or mailings of field reports further ensures they will be done daily, in a timely manner. The daily field report is typically filled out by the site superintendent and signed by the project manager if that is an on-site position. In the absence of a site superintendent on smaller projects, the report must be filled out by the responsible project manager on a daily basis. Field reports need to include information listed in these pertinent areas:

- **Title Box**. Indicate project name, company job number, project location, and the name of the individual completing the report.

- **Date/Page**. Account for each page of the total report. Use as many report forms as necessary for a complete record of the day's events.

- **Weather**. Include a short comment on the typical condition of the day (rainy, cloudy, sunny, etc.). Record the temperature at the beginning and the end of each day.

- **Visitors.** List all visitors to the job site that day who are not part of the regular workforce. Include a brief remark as to the reason for the visit. Note any special situations deserving immediate attention. Include references to appropriate documentation.

- **Work Force**. Indicate the number of each type of employment classification for each type of labor involved.

- **Activity Performed**. Directly adjacent to the respective labor force information, include a short description of the type of work performed.

- **Equipment**. List all major pieces of equipment on the site, whether or not they arrived on that day. Include brief descriptions of equipment work performed.

- **Items Received/Sent**. List all materials and equipment received or sent that day only.

- **Location**. Provide sufficient references to allow accurate location of the work. For example, "Continue foundation wall forming along column line A, between lines 4 and 6."

- **C.O. No.** If the work described under "Location" applies to a change or a proposed change, insert the number of the change order or change estimate number in the column provided.

- **Problems/Comments**. Self-explanatory.

Change Orders

Contract law does not allow the prime or subcontractors to make any changes to an existing contract. An addendum in specified work requires the contractor to write up a contract change order and then seek written approval from the owner. The owner is not obligated to sign a change order, but the owner's approval is absolutely necessary to avoid litigation problems. The owner cannot be forced to pay for work that was not authorized. Contract law specifies that "all changes to the original contract must be in writing and signed by both the property owner and contractor." Specify that ordering materials shall be done after the contractor reviews plans and specifications. Visiting a job site is not sufficient to see all of the details of the job. During project operations, if materials or items are ordered on an "as needed" basis, delays will occur if the materials or items are unavailable or on back order.

Cost-Loaded CPM

Cost-loaded CPM comes into use with unit-price contracts. These contracts use prices that are based upon an itemized breakout list of all work activities involved in the contract. Each is assigned a unit price. Using this method, the final line-item price on each activity item is computed by multiplying the unit price times the item quantities. The unit price is then used as the controlling price for each item. Cost-loaded CPM is built from the unit-price bid sheets supplied in the original contract by the activities subcontractors. This is used by the scheduler as a schedule of values, and must not be confused with the lump-sum method. In a lump-sum project, the contractors contract to do all the work defined in the plans and specifications for a fixed price stated in the contract.

Any quantity variations that are not added by the project team as change orders, are not used in factoring. To cover these possible variations, a schedule of values is also requested from the contractors in their bids. A schedule of values is basically a unit-price bid with some notable differences. For example, item prices are shown for ease in making monthly progress payments only; they cannot be used to price change orders and have no contractual obligations. Cost-loaded CPM is used to advance phased scheduling on progressive milestones.

A common problem that occurs when owners request cost-loaded CPM is their assumption that cost-loaded CPM will always result in easier progress payments management. This is true of lump-sum projects, but the situation is reversed in unit-price contract projects, because a cost-loaded network does not synchronize with the pay line items on unit-price contracts.

Therefore, in a lump-sum contract with cost-loaded CPM, the unit-price breakout should be submitted after the contract approval but before the first progress payment. In a lump-sum contract without cost-loaded CPM, a schedule of values should be submitted after the contract approval but before the commencement of work.

Typically, all engineering projects are based on a unit-price contract where the actual unit prices stated in the original bid will be held as the factor for all progress payment disbursement amounts. Because a unit-price contract does not have a fixed ceiling price, the final cost to the owner will be determined by the quantities actually completed. To adequately protect the financial investment of the owner, the scheduler must be careful to determine that the field quantities are precise. Conversely, in a lump-sum contract project, if an error is made within the first month of work it can be compensated for in a later progress payment. The contractor cannot receive more money than the stated fixed price, even if some of the final quantities varied from the anticipated amounts.

In a lump-sum contract, the contractor agrees to build the complete project as per plans and specifications. Anything necessary to accomplish the completion of the project must be considered part of the contract, even if not specifically stated in those plans and specifications. Conversely, in a unit-price contract the contractor may still be required to construct a complete functional project, but if any variation occurs in the quantities of any of the separate bid items listed, the contractor is entitled to an amount of money equal to the unit price of the bid item, multiplied by the actual quantity that was supplied or installed of that item.

Sorts

Sorts are the summary reports of all estimated, projected and actual information current in the project as of each report's run date, and they must be organized and distributed during project operations. The computerized network schedule has sorts that are interlinked with one another and provide summary reports that must be "sorted" out to the appropriate project team members, owner, and responsible activity and procurement personnel by the type of information needed by the reader. The large menu of output sorts is another big advantage of a computerized CPM schedule. Sorts allow the various members of the project team to receive summary reports of progress, pending construction, or future scheduling in the output sort best suited for their needs. Most CPM programs will yield a sort menu as follows:

1. Sort by activities
2. Sort by *i-j* numbers
3. Sort by early starts
4. Sort by job logic
5. Sort by project milestone
6. Bar chart printout
7. Human resource leveling
8. Key-milestone-date sort
9. Critical item sort
10. Sort by events
11. Free float per activity with total float summary
12. Critical-path sort
13. Critical-equipment sort
14. Sort by total float/late start
15. Limited periodic look-ahead sorts
16. Schedule of anticipated earnings
17. Cost by activity number

Most project managers and site supervisors find the sorts by total float and by milestone most valuable for their project management needs. The total-float sort starts with the low-float (most critical) work activities listed first for immediate attention. The less critical high-float items show up later on the list. By using the periodic look-ahead sorts, one can also home in on specific time periods. A thirty, sixty, or ninety day look-ahead sort will list only those critical items that will occur in the next thirty, sixty, or ninety days. These sorts are extremely useful for the field superintendent, area supervisors, and field schedulers.

Other members of the project team need other types of data sorts to make their work more productive. For example, materials-control people find the critical item sort more convenient in tracking their required delivery dates than extracting those dates from a milestone chart. As the revised delivery data and actual progress are fed into the computer, revised printouts quickly reflect the delivery changes and their effect on the field schedule. Field people usually find that the key-milestone-date sort better suits their needs. That section of the CPM schedule is the basic document that the field scheduler uses to make the detailed weekly work schedules in the field.

The rapid turnaround of data sorts by the computer also allows the project team to perform what-if exercises with the logic diagram. When scheduling problems arise, the project team can try alternative solutions by reworking elapsed times for problem activities. This generates new early and late start dates that can be shifted to improve the critical path. The computer calculates a new critical path in a matter of seconds, with immediate access to the new output data sorts on the computer monitor.

If it is desired to determine all items that can be started as of any given date, reference to this sort will show under the column headed "Event Start," subhead "Early," the earliest date that any activity can be begun. On-the-schedule reports show that the effect of using total float for a given activity can be shown. Activities that are in series are consecutive operations, a fact that can be noted on both the project operations printouts and the detailed CPM network plan. By altering the events of other activities in the network, days of slack can be realized for the activities not on the critical path schedule. Zero free float in activities red-flags the scheduler that this activity is heading into negative float, which in turn robs total float from the entire project.

As-Planned, As-Built, and Adjusted Schedules

Construction schedules are also used in the context of justifying time extensions and delays in project operations. The kind of analysis that can accomplish these demonstrations are either performed before the situation is actually encountered in a claim situation, or immediately afterward. In the first case, reasoning is presented to convince the owner that a potential change will affect the schedule by a precise number of days. This is accomplished with the use of two separate but related schedules. The first is the as-planned schedule. This is the schedule that indicates the new sequence of work, incorporating all the influences of the change along with the ultimate effect on the project end date. That end date change is attributable to the change.

The second situation is the one in which construction schedules must be used to demonstrate where you would have been had it not been for the subject change. It involves an "ex post facto" analysis, often in a delay claim situation. This particular kind of analysis is most useful in those cases where the project has been affected by a change, but the usually complex interaction that makes up the total project history confuses any clear cause-effect relationships. The dilemma is solved with the use of three schedules. First, again, is the as-planned schedule. Next is the historical as-built schedule that through its periodic updates incorporates all the complicated effects on the construction sequence. Finally is the adjusted schedule. This schedule reconstructs each update as necessary, removing the effects of the change of interest. The result is a schedule that includes all the other effects without the effects of the change being considered. The difference in the end dates between the as-built and adjusted schedules becomes the amount of time attributable to the single change.

Chapter 15

Software Program

Installation

This software program is configured to run on the PC spreadsheet programs Lotus™ or Quattro™, versions 3.0 or higher, on a Windows™ platform, version 3.1 or higher. Boot up the computer, open the spreadsheet program, insert the disk and switch to that drive to begin using the program, or install the disk to your hard drive and run program from there. This latter method will produce greater speed in compilations of summary sorts because operating memory will not be restricted to RAM. For the purposes of clarity in instruction within this book, commands are shown in text as bold with underlining. The underlining represents the command keys if you are entering commands through the keyboard, or the bolded letter you would point to and click if you are using a mouse. Link absolute cells and macro spreadsheets according to your spreadsheet application's manual.

Cell ranges within each sort that have data in italics are for your company and project data input. Areas of format layout are protected against overwriting, as is the tabulation programming. The programmers endeavored to make the data entry process bulletproof, because I make mistakes just like everybody else. I still wipe out cells of work from going too fast and overrunning a range select inadvertently during data transfer. You know what I'm talking about... When you do something so stupid, you look around the room to see if anyone saw you acting dumber than a box of rocks. So ***FastPro*** is designed with protected cells that will keep the master sort formats from being erased accidently. Only your data input cells can be messed up, and you're not going to lose the master by experimenting with the program spreadsheets, so relax. Do some playing with it and see how your activities integrate through the different sorts.

Data Entry

To enter data in any sort spreadsheet, move to the cell in the spreadsheet you wish to input the information to and type in the data. Press Enter. As you type, the entry appears on the edit line. As you press Enter, the data entry inputs to the current cell. If you enter data in a cell that already had information in it, the old data will be deleted and replaced with the new input data. If you need to input data into more than one cell address at a time, you do not need to repeat the data entry process in each cell individually. Instead to increase the speed of multiple data entry you can select blocks of cells, called ranges, make information or layout changes, then press Enter. The selected range will accept the command. This is called a *range command,* because it affects the entire selected range within the spreadsheet. By selecting groups of ranges, you can step up to the macroactivity of data entry, called *global commands*, which are data entry commands which affect the entire spreadsheet at once.

You can press a directional key such as Tab, Page Down, Page Up, or Arrow, and move to the data entry cell with one keystroke, enter data, then lock in the entry for automatic computation by pressing Enter. Activities are laid out in the vertical axis of columns and number figures are laid out in the horizontal axis of rows. Each row and column is integrated and the two or more data entries made on each activity will automatically compute to the subtotal cell address in each column. These subtotal column cell addresses are further integrated to the master total cells for total computation of all activities in each phase. These phases are preset as you see them, in a typical commercial project sequence. For residential projects, simply use the activities that are relevant to your project and leave the others blank. Data will automatically wash through the blank cells. You may add or delete activities as you wish; however, it is recommended that you save your new spreadsheet sorts under a new file or to a separate disk, leaving the master format for future projects.

This software is formatted in WYSIWYG (What You See Is What You Get), and the data that appears as you enter it is what the finished and printed product will look like. You can create two kinds of cell entries for different kinds of data entry: *values* for the number figures and *labels* for your word processing data. A label is configured solely as a text entry, whereas a value is a number or formula used by the computer for computation. The computer recognizes which type of data entry is being used in the cell by the first character in the entry. If the character is a number, the computer will use the data for numerical computation factoring. If the first character is a letter, the computer then recognizes the entry as a label. You can interlink text entry activity names with their relative values by the use of activity numbers. You then further link those numbers globally in the spreadsheet in groups of phases by their *i-j* numbers. This in turn produces the sequential linkage necessary for the computer to produce the CPM program logic.

Labels make the numbers and formulas in the spreadsheet anchor the related data specifically to that activity. As a text entry, a label can be a string of up to 512 characters or bytes. Labels include headings, titles, explanations, and notes; basically, any word processing that will explain, clarify or add information to the description of that activity. When you enter a label, the software program adds that data entry to the relative cell address and the computer then recognizes that cell as a label. When you enter a label, the program, by default, will left-align your entry in the cell. You can change the format of the label by typing one of the following label character prefixes before typing the data:

' Left-aligned (This is the default setting)
" Right-aligned
^ Centered
\ Repeating

A vertical bar (|) entered as a label prefix notifies the computer not to print the contents of the cell. Column headings should be aligned with the data in the columns. This provides instantaneous orientation to the human eye of data with text. When an entry fills the cell width, the entry can be centered, right-aligned, or left to the default setting of left-aligned. Standard spreadsheet format requires that labels be centered when the column heading is narrower than the data in the column. Repeating labels will fill the width of the cell and, if you change the width of the column, will automatically repeat more to refill the new column width. To use a label prefix as the first character of a label, first type the label prefix if you wish to change to from the default setting of left-aligned. If no change in label alignment is desired, simply type in the data and the computer will automatically align the data with a left justification.

If the cell entry data is numeric, you must first type a label character as a prefix. This is done because, without a label prefix, the computer will mistake the entry as a number or formula for computation and will begin to compute the data as soon as you hit **E**nter. The classic example of this data entry technique is a telephone number. If we were to enter the phone number 555-702-1995, the computer would instantly mistake this data entry as a mathematical formula and begin subtracting 1995 from 702, and the remainder then would be subtracted from 555. We would end up with a bizarre negative number in a cell where the client's phone number was supposed to appear. Computers are, after all, just dumb boxes that record data. The program cannot think for you or anticipate your intentions, and it is restricted to observing whatever format the program is configured with in dealing with your data entry. A simple oversight of label prefixes or typographical errors will result in what is called a *syntax error*, which is the inability of the computer to read your data, or even worse, sending that information off to do something other than what you had intended.

Another classic example I use in my classes, that serves as an illustration of this data entry fundamental, is entry of dates. If we were to enter 7-15-95 as the date, we might recognize that as a date but the computer recognizes that entry, again, as a mathematical formula and, instead of the date appearing in the cell, the answer resulting from subtracting 95 from 15 from 7 would appear. Therefore, to retain a numerical entry as such and not signal the computer to use the entry for computation, you must use a label character as a prefix. A telephone number would be preceded by an opening parenthesis followed by a closing parenthesis at the end (555-702-1995). Here the opening parenthesis serves as a label character before the figures and signals the computer to treat this data as a label and not to use it for computation functions. Similarly, a date entry can be handled with either of two techniques. The first is by using the @Date function of the software program; the second is by entering the date data as 7/15/95, which the computer recognizes as a date label and not a formula function.

If a label is accidentally evaluated as a formula, press F2, (**E**dit), then press **H**ome, then type the label prefix and press **E**nter. Should the program place you in the edit mode, press **H**ome, type the label prefix at the formatted insertion point, and press **E**nter. If the label characters string is longer than the cell's width, the label will automatically scroll across and through the cells to the right, if there is no data in those cells. If data is already in those cells, the program will cut off the display of the current cell data entry at the right border of that cell. The computer, however, still stores the complete data string and will display the entire cell entry on the current edit line when the cell is selected. To display the entire label in the spreadsheet, you can insert new columns to the right of the current cell containing the long label, or you can widen the column sufficiently to contain the entire label characters string.

Entering Numbers

To enter a valid number in the spreadsheet, you can type any of the standard numerical characters (0 through 9) and certain other characters that will serve as formula prefixes. The following are the numerical entry procedures for this software configuration:

- You can start the numerical entry with a plus sign (+).
  The plus sign is not stored when you press **E**nter.
  +555 is stored and displayed as 555.
- You can start the numerical entry with a minus sign (-).
  The number is stored as a negative number.
  555 is stored and displayed as -555.
- You can include one decimal point.
  .555 is stored as displayed as 0.555.
- You can place the numerical entry in parentheses.
  The number is stored as a negative number. The computer will automatically drop the parentheses and install a minus sign prefix to the number.

(555) is stored as displayed as -555.

- You can begin the number with a dollar sign ($).
  Unless the cell is formatted as Automatic or Currency, the dollar sign will be dropped and will not be stored to memory when you press Enter.
  $555 is stored and displayed as 555 in an unformatted cell. If the cell is formatted in either Automatic or Currency, $555 is stored and displayed as $555 (with no decimal places unless you also format the cell to display decimal places).
- Three digits must follow each comma you include.
  Unless you format the cell as Automatic or Comma, commas are not stored in memory when you press Enter.
  123,456 is stored and displayed as 123456 in an unformatted cell.
  123,456 is stored as 123456 and displayed as 123,456 in a cell formatted as Automatic or Comma (with 0 decimal places).
- You can enter the number in scientific notation.
  A number is stored as scientific notation only if it requires more than 20 digits.
  555E3 is stored and displayed as 555000. 5.55E30 is stored as 5.55E+30 and displayed as 5.5E+30. 555E-4 is stored and displayed as 0.0555. 5.55E-40 is stored as 5.55E-30 and displayed as 5.5E-40.
- You can type the number with a percent sign (%) at the end.
  Unless the cell is formatted as Automatic or Percent, the computer will divide the number by 100 and the percent sign is dropped.
  555% is stored and displayed as 5.55 in an unformatted cell. 555% is stored as 5.55 and displayed as 555% in a cell formatted as Automatic or Percent (with 0 decimal places).
- You can enter a number with more than 18 digits.
  When the computer has read 18 digits, the number is rounded off and ends with one or more zeros.
  123456789987654321123456 is stored as 12345678998765432100 and is displayed as 1.2E+19.

If the number is too long to display in the cell, the computer will display as much of the number as will fit within the cell borders. If the cell data input is done using the General default setting, and if the integer part of the number does not fit into the cell, the program will display the number in scientific notation. If the cell's data entry is with a format other than General or if the cell is too narrow for the number to be displayed in scientific notation, the number will be stored in its entirety but will be displayed as asterisks.

Entering Formulas

The real power in this program comes from its formula computation capability. Formulas are integrated through the sorts to provide computerized interlinkage for assembling CPM data. You enter the numbers and formulas in the sorts spreadsheets and the CPM program calculates the results of all the formulas. The real beauty of this type of system is that, as you add or change data, you do not need to recalculate the effects of the changes throughout the entire spreadsheet the program will automatically do this for you. This is how we run "What if" scenarios, by changing a factor here or there to see what the end results would be if certain changes were undertaken without upsetting or revising the entire spreadsheet. If the changes are acceptable, they can be introduced into the specific sort you are working in or globally to the entire network schedule.

You can enter formulas that operate on labels, numbers, and other cells in the spreadsheet. Like a label, formulas can contain up to 512 characters. A formula can include text, numbers, operators, cell and range addresses, range names, and functions. A formula cannot contain spaces except within a range name or text string. You can create four types of formulas: numeric, string, logical, and function. Numeric formulas work with numbers, other numeric formulas, or string functions. Logic formulas are true or false tests for numeric or string values. Formulas can operate on numbers in cells. The formula 2+10 uses the calculator within your program to compute the answer. A much more useful formula involves using cell references in the calculation. The current cell shows the formula and, by interlinking the formula to different cell addresses, the spreadsheet shows the result of the calculations throughout all the assigned cells. The resulting computations will change if you change any number in the other cells. This instant recalculation is the real power behind CPM programs, because it allows you to change paths and their related data on the computer first, to analyze the moves before actually making them in the field production.

Formulas consist mainly of operators and cell references. You can type each, or you use a better way to enter the cell addresses. When the program requires a cell address as a source or target reference cell, use the directional keys to move to the cell or type the cell address, then press **E**nter. The preferred method is to move to the cell directly as opposed to typing the cell address, because a simple mistake in typing the cell address will result in that data's going to the wrong cell. Thus one error will lead to multiple errors when the data is calculated through to other cells. The error will not show up on the display and the mistakes carried throughout calculations within the master spreadsheet may not be caught by the scheduler. Errors of this nature are common and will show up when the schedule is cooked, in number totals that seem unreasonable compared to estimated figures or historical data. Backtracking the error usually confirms that the data entry was being sent to a wrong cell address.

Formulas that interlink between files are usually very long and complicated. Accordingly, standard operating procedure is that the target file you are transferring data to should be open behind the current file you're working in. This allows you to specify the target file by selecting cells or ranges instead of typing the cell or range address (and thereby incurring the risk of error from a typo in data entry). In formulas, data entry can be done in one of two ways. A cell address in a formula is known as a *cell reference.* Usually, when you use the Edit Copy command to copy a formula from one cell to another, the cell references are adjusted by default. If you copy the cell formula in your source cell, once transferred, the cell references change by default in your target cell(s). This adjustment is known as *relative addressing*.

An absolute address in a formula, on the other hand, does not change when you copy the formula to another cell. You specify an absolute address in a formula by typing a dollar sign ($) before the column and row addresses. For example, +$A:$A$1 is an absolute address. If this address were in cell B5 and you copied it to cell G10, the cell reference would still be +$A:$A$1. To specify an absolute cell address in the select mode, press F4 (Abs). In addition to using relative and absolute cell addresses, you can use mixed cell addresses.

In a mixed cell address, part of the cell address is relative and part of the address is absolute. If we were to change our above example to read +A$1, this would now be a mixed address because one factor is relative (+A), and the other factor ($1) is fixed as an absolute address by the use of the dollar sign. Whether a cell reference is relative, absolute, or mixed has no effect on how the formula is calculated by the computer. The computer will differentiate between them when you copy the formula from the source cell address to the target cell address.

Logic Formulas

Logic formulas are true/false computations based on the formula within the cell. A logical formula compares two values and returns the number 1 if the formula computes to be true, and 0 if the formula computes to be false. This kind of formula is used mainly in establishing database criteria ranges. As an example, Table 15-1 shows typical logic formulas, how the computer reads them, and how the program then computes them. Note the difference in results from similar formulas simply from the syntax of data entry. This points out to the project scheduler the importance of avoiding typos in data entry. The trick to accurate data entry is to take time to double check your numbers twice.

Formula	Computation	Result
5+3*2	(5+(3*2))	11
(5+3)*2	(5+3)*2	16
5+4*8/4-3	5+(4*(8/4))-3	10
5+4*8/(4-3)	5+((4*8)/(4-3))	37
(5+4)*8/(4-3)	(5+4)*8/(4-3)	72
(5+4)*8/4-3	(5+4)*(8/4)-3	15
5+3*4^2/6-2*3^4	5+(3*(4^2)/6)(2*3^4))	-149
-3^2*2	-3(3^2)*2	-18
-3^(2*2)	-(3^(2*2))	-32

Table 15-1
Logic Formulas

Using String Operators

A formula is a mathematical instruction to the computer to perform a calculation or series of calculations. We use *operators* to specify the calculations to be performed, and their order of precedence or sequence. ***FastPro CPM Scheduler*** uses operators in numeric, string, and logical formulas. Just as ranges are the macroactivity of cells, operators are the macroactivity of formulas. You begin to see now how our building blocks work. We are structuring circles within circles. In this way, our CPM schedule will have all levels of data from the minute tasks and work items right on up to the major phases of activities, which gives the computer the factoring linkage necessary for program logic. Time manipulation is achieved in the network schedule through the use of tools such as string operators.

Numeric operators are used for the four basic mathematical functions of addition, subtraction, multiplication, and division, as well as *exponential factoring* (raising a number to a power). The simplest numeric formula uses just the plus sign (+) to repeat the value in another cell. The most complicated can involve all the above four standard numeric functions, as well as exponential factoring of any or all integers in the formula, in differing sequence or precedence, all together or individually. This is a very powerful function that provides fast, error-free data quicker than any human being could.

To transfer string operator formulas from source cells to target cells or ranges, you can select the target cells or ranges (which is the recommended method) or you can enter the following commands by using doubleangle brackets:

+<<Source range>>A:B3+<<Target range>>

Any length of formula or string command can appear where the example formula of A:B3+ appears. If the data transfer is to go to another sort, you must include the entire command path, which is fundamental in DOS operations. For example:

+<<C:\FastPro\data\target file.doc>>.

If a formula is entered using an operator shown in table 15-2, the program calculates the data by following the order of precedence that is noted in the table.

Operator	Operation	Precedence
10^{10}	Exponentiation	1
- (prefix)	Negative value	2
+ (prefix)	Positive value	2
*	Multiplication	3
/	Division	3
+	Addition	4
-	Subtraction	4
< >	Less than, greater than	5
< =	Less than or equal to	5
> =	Greater than or equal to	5
=	Equal to	5
#NOT#	Logical NOT	6
#AND#	Logical AND	7
#OR#	Logical OR	7
&	String formula	7

Table 15-2
String Operators

Using operators in string formulas differs from regular numeric formula entry in that a string is a label or a string formula. Only two string formulas exist. You can repeat another string, or you can combine two or more strings. The simplest string formula uses only the plus sign (+) to repeat the string in another cell. The formula to repeat a string cell is the same as repeating a numeric cell. A formula is considered a numeric formula because the formulas refer to cells that contain numbers. A string operator is considered a string formula because the formulas refer to cells that contain string formulas. The operator for combining strings is the ampersand (&).

The first operator in a string formula is the plus sign, then all sequential operators in the formula must be ampersands. A cell containing a label is read by the computer as having a value of zero. If you do not use the ampersand but instead use any of the numeric operators, the computer will compute the data as a numeric formula. By inserting an ampersand, the computer now treats the data as a string formula. If you insert any other numeric operator after the prefix plus sign at the beginning of the entry, the formula will result in a syntax error and ERR will be displayed. The string formulas +A1&B2+C3 and +A1+B2&C3 will both result in syntax error due to the additional numeric operator inserted in the formula. If you make an error during data entry that will not compute through the system, an audible beep will come from your CPU, which is signaling you that an invalid formula has been input. Common errors that make a formula invalid are extra or missing parentheses in the formula where the program encountered an error, incorrect precedence or arguments in the formula, and misspelled function names (syntax error). The following is a list of the most commonly encountered types of scheduling data entry errors:

+A1/(2-A3	Missing right parenthesis
@SIM(A1..A6)	Misspelled @SUM function
@IF(A1>200,200)	Missing argument in function

Menu Commands

To issue a command with the keyboard, you activate the main menu by pressing the Alt key or the Menu (F10) key. Use the arrow keys to move the menu pointer to the name of the command you wish to select, and then press Enter; or type the underlined letter of the menu option. If you are using a GUI system with a mouse, just point to the command you wish to choose and click the mouse which serves as Enter. This GUI command automatically activates the menu and chooses the command. As you choose a menu option from the main menu, the next set of commands appears in a pull-down menu. All of the main menu commands lead to pull-down menus, which provide further detailed task commands within the larger main menu option which holds the main directories.

In Windows™ applications, if the pull-down menu option has another level of commands the commands appear in a *cascade menu*. Pull-down menu options that lead to a cascade menu are usually followed by a solid triangular marker, depending on which type of processing program your system is running on. If you make a mistake while choosing menu commands, press **E**sc to return to the preceding menu. If you press **E**sc at the main menu you deactivate the menu and return to the **R**eady mode. When using menus, pressing **C**trl+**B**reak is equivalent to using the **E**sc key.

When you make changes to the master format provided, all the work exists only in the computer's working memory (RAM). If you do not save new spreadsheets or changes before you quit the program, you will lose all work done. Before exiting the program, save all new work or changes in a separate file or save to C, your e hard disk, or to a backup disk in either the A or B drive. If you save your changes on the master format, the information existing previous to your changes will be unrecoverable. This will be another of those outstanding moments in computer applications history when you'll want to drag your computer outside and shoot it right between the running lights.

Saving Your Work

When you have created a sort on the master format, the data and information will have to be saved on either your hard disk or a floppy disk. When a sort is keyed in for the first time and is displayed on the screen, it is in RAM and is only temporary. If you turn the computer off or if the power goes off, you will lose the data and information and have to re-enter it. Only when you save a sort to a disk is it saved permanently. Every time you load ***FastPro*** on your spreadsheet program, you will be able to bring up a saved sort back to the screen.

To save a sort you must execute a save command. ***FastPro*** provides four methods for doing this. They are:

1. Using the toolbar
2. Using the menu bar with the mouse
3. Using the menu bar with the keyboard
4. Using speed commands

FastPro contains two commands that are used to save sorts. One is the **S**ave **A**s command and the other is the **S**ave command. When you begin creating a sort it does not yet have a file name. You assign it a name when you save the file. The first time you save the file, you may choose either the Save or Save As command. ***FastPro*** gives you the opportunity to assign the file a name of your choosing. After the first time you save a sort, however, you should use the Save command to continue saving changes to your sort. ***FastPro*** automatically uses the same file name when the sort is saved by using the Save command. The Save As command is used to save changes to an existing file under a different file name.

Speed commands are executed by pressing either one or two keys on the keyboard in conjunction with a function key, or by pressing a function key alone. In some cases, there is more than one speed command for the same function. For example, to save a sort file for the first time you can press either Shift + F12 or Alt + Shift + F2. Either method executes the Save command.

To save a file using toolbar, complete the following steps:

1) Click on the Save icon.
2) At the Save As dialog box, key in the name of the sort.
3) Click on OK or press Enter.
4) At the Summary Info dialog box, click on OK or press Enter.

To save a sort file using the menu bar with the mouse, complete the following steps:

1. Click on File, then Save.
2. At the Save As dialog box, key in the name of the file.
3. Click on OK or press Enter.
4. At the Summary Info dialog box, click on OK or press Enter.

To save a sort file using the menu bar with the keyboard, complete the following steps:

1. Press the Alt key, key in the letter **F** for File, then the letter **S** for Save.
2. At the Save As dialog box, key in the name of the file.
3. Click on OK or press Enter.
4. At the Summary Info dialog box, click on OK or press Enter.

Procedures

To begin, enter into the Activities Sort and begin building your schedule by adding activities and creating links between them. Add an activity, fill in its duration and other details in the activity form, and connect the activity to its predecessors and successors. Create activities' linkages by connecting set of dates (early and late start and finish events). Compare today's schedule dates to the plan for each activity by varying their events and end points for completion. Switch to each sort and enter data to define the data columns that appear in each sort, along with specific timescale data that is programmed into the rows. This configuration will give you computation factoring on all the data on each activity and work item.

Once data entry is complete, begin organizing and summarizing the sorts. Activities that share a common attribute, such as a subcontractor or responsibility, can be grouped into collective blocks known as *ranges*. Activities can be grouped into bands based on activity codes, such as project team member responsibility or related assigned subcontractors. You can also group activities by resource, cost accounts, dates, calendar milestones, or the work breakdown structure of the project.

To integrate data from other applications, use Object Linking and Embedding (**OLE**) within that application to originate the transfer. Use source application attachment tools to link text, graphics, spreadsheets, and drawings. Plan resource usage and financial budgets for the duration of the project by allocating, analyzing, and tracking resources and costs by activity. Sort by activities costs will show when required resources are exceeding normal availability, and another column will show when required resources are exceeding maximum availability.

Changing Data

After you have built the data within the spreadsheet, you may want to change the data in specific cell locations or input new data for a new project. You can change an existing entry in two ways. First, you can replace the entire contents of a cell by typing in a new entry and pressing Enter. Second, you can change part of a cell's contents by selectively editing only the portion of the data that you wish to change, by inserting the cursor ahead of the data to be changed. Make the changes, delete the old data, and press Enter to store and display the new data. To replace the entire contents of a cell, move to the cell you want to change and press Enter to select that cell. You can either press Delete to empty the cell and input the new data or simply type the new data, and press Enter to store and display the new data. To selectively edit certain portions of a formula or string within a cell, move the cursor to the cell and press F2 (Edit) to enter the edit mode.

If you press Esc while in the Edit mode, you clear the edit area. If you press Enter in a blank edit area, you do not erase the cell contents, and you return to the Edit mode. Should you make an error in your data changes, you can reverse the changes to the original data by the Undo function (Alt+Backspace), or select the Edit Undo command to return the original data to the cell. If you type over an existing entry you can undo the new entry and restore the old one. The Undo function undoes only the last action performed, whether it involved entering data, using a command, to completion or running a macro. When Undo is enabled, the computer must retain the most recent action in RAM. This feature requires a great deal of computer memory. Accordingly, the bigger your RAM the faster the command entries will respond.

How much memory is used up in RAM depends on the different actions involved. If you run low on memory, you can disable the Undo function by selecting Tools/User Setup and deselecting the Enable/Edit Undo feature. If you run out of memory while the program is undoing an action, the program will suspend the operations and save the data currently in RAM. The Undo function is very powerful and also tricky, so you must use this function carefully. To use Undo properly, you must understand what the computer considers a change. A change occurs between the time when the program is in the ready mode, data is entered, and the next time the program is in the ready mode.

This is the extent of data entry change that will reside in RAM for recovery. Let's take a practical example to illustrate this point. Suppose that you press F2 (Edit) to go into the Edit mode to change a cell. You can make any number of changes in the cell, press Enter to save and display the changes, and return to the Ready mode. If you were to press Undo at this point, the computer would return the spreadsheet to the original data condition the spreadsheet was in during the last Ready mode. The cell returns to the previous data existing before you performed the edit. You can change a range or group of cells at one time or even erase everything in RAM with just one command. The Undo command is powerful enough to cancel all the effects of the command.

On the other side of the coin, some functions cannot be canceled, including the Undo command. Undo, when used by mistake, cannot be canceled. If you press Undo (Alt + Backspace) at the wrong time in data entry and cancel an entry, you cannot recover the entry. If you are not in the habit of saving your work every fifteen minutes, then this will be the first time you've lost four hours of work on a pilot error glitch. Really stinks, huh? Wait, wait! It gets worse! Just wait until it dawns on you that you've another four hours to go to reconstruct your work all over again. Then you're going to go ballistic. Moral of story: Hit the Save button every fifteen minutes and discipline yourself to do it automatically, forever. Fifteen minutes of lost work can be caught up on with only a couple of choice swear words, instead of missing your dinner for another four hours in a manic-depressive state.

Other commands that cannot be canceled include:

- The Graph commands (although you can cancel these commands by using the Edit commands in the Graph tool)
- All commands on the Control menus
- All commands that affect an outside source but have no effect on the current spreadsheet file
- Actions that move the cell pointer or scroll the spreadsheet, including **GoTo** (F5) and **Window** (F6)
- Formula recalculations that result when you press Calc (F9)

Database Terminology

A *data table* is an on-screen view of information in a column format, with the field names at the top. A data table contains the results of a Data What-if Table command, plus some or all of the information that was used to generate the results.

- A *data table range* is a spreadsheet range that contains a data table.

- A *variable* is a formula component whose value can change.

- An *input cell* is a spreadsheet cell used by the computer for temporary storage during calculation of a data table. One input cell is required for each variable in the data table formula. The cell addresses of the formula variables are the same as the input cells. An input cell can be a blank cell anywhere on the spreadsheet. The preferred procedure is to identify the input cell by entering its title with an appropriate label either above the input cell or immediately to the left.

- An *input value* is a specific value the computer uses for a variable during the data table calculations.

- The *results area* is the portion of the data table where the calculation results are placed. One result is generated for each combination of input values.

The results area of a data table must be unprotected, or the data will not write to the area. The formulas used in data tables can contain values, strings, cell addresses, and functions. You should not use logic formulas in a results area, because this type of formula always evaluates to either 0 or 1. Although using a logic formula in a data table does not cause computer error, the results are usually meaningless.

Using Databases

A database is a directory of related information, organized so that the end user can sort, list out or search through the data to find desired information. A database may contain any information pertaining to the project, grouped into whatever divisions make the data retrieval more convenient for the end user. In the software program, the computer recognizes a database as a range of cells that spans at least one column and more than one row. Because a database is a data list, its manner of configuration and organization of data make it distinct from ordinary label cells. Just as an information list must be organized to be useful, a database must be organized for the computer to access the data within the database parameters.

This software program has three types of database organization. The simplest database organization is a single database contained within a single spreadsheet sort. In real-world applications, this is the most common and frequently used type of database organization. The next step is organizing groups of multiple databases from these single database organizations. This software uses program logic to integrate multiple single database organizations with others occupying a different portion of the same spreadsheet sort. This second level of database organization can be precedence-linked or non-sequential, according to your needs in programming the data retrieval. The last type of database organization involves integrating multiple databases in two or more spreadsheet sorts. Remember, however, that a single database table cannot span different spreadsheet levels. In order to accomplish this third type of database organization, you must relate databases that are different spreadsheet levels, and the computer will then interlink them to produce a more efficient overall database structure.

Any of these database organizations is similar to any other group of label cells. The computer separates them from label addresses by specifying the two components of a database system, *fields* and *records*. A field is the smallest unit in a database, made up of a single data item (datum). An example would be a client record that includes separate fields containing a company name, phone number, address, and contact person. A record is a collection of related fields that build an information group. In our example, the four fields noted for each company, when accumulated, would make up one record for one company. In simpler terms, the computer recognizes a field as one cell address and a record as a range of cell addresses within the database.

The functional purpose of building a database is to organize information to expedite data retrieval for the end user. In operating with a database, data retrieval involves searching in *key fields*. A database key field is any field on which you base a list, sort, or search operation. In a key field search operation you can use a search operator, such as an address, as a key field to sort the data in the company database and to assign the information retrieval according to geographic areas.

It is important that you do not intermix database tables with data tables, as this will result in computer data search syntax error. The computer will compute data tables through its precedence configuration, which is the menu function **D**ata **W**hat-if **T**able command. In database tables, however, the computer searches multiple ranges used with relational databases, because a single database may contain several tables of related information.

Building a Database

You can create a database as a new spreadsheet file or as part of one of the master sorts existing within the program. If you decide to build a new database as part of an existing spreadsheet, be certain there is sufficient room to store and retrieve a spreadsheet database which resides within the spreadsheet's row-and-column format. Labels or field names that describe the data items contained within the record or field appear as column headings. Information about each specific data item (field) is entered in a cell in the appropriate column.

Theoretically, the maximum number of records you can input in a database corresponds to the maximum number of rows in the spreadsheet sort. This program is configured with 8,191 rows for your benefit in scheduling larger commercial projects. Residential projects will usually not need that depth of database volume, but this software was programmed with that data entry ability in case you should need it. Realistically, however, the number of records in a specific database is limited by the amount of memory in your computer. The actual amount of memory equals the internal working memory (RAM) plus disk storage for virtual memory, less the room needed within the database to hold data extracted by the **D**ata **Q**uery commands. When you estimate the maximum database size you can use on your computer equipment, be sure to include enough blank rows to accommodate the maximum output you expect from data extract operations. A working technique I use on small systems that run short of memory is to split a large database into separate database tables on different spreadsheet levels, if all the data does not have to be sorted or searched as a unit. For example, you may be able to separate a telephone list database by name such as A through M in one file and N through Z in another file, or perhaps sort by area codes.

Create a database as a new spreadsheet or as part of an existing spreadsheet. If you decide to build a database as part of an existing spreadsheet, chose a sort area that is not needed for other data. This area needs to be large enough to accommodate the number of records you plan to enter during the current data entry session and for future data entry in that sort. When interlinking separate sort spreadsheets, make sure that the spreadsheets do not interfere with one another with respect to available target file room for data. The easiest way of doing this is to use the your **G**lobal **S**ettings command to make certain **G**roup mode is not selected. This will prevent the data from overwriting the existing data inadvertently. This also prevents the column width and row height settings, as well as the insertion and deletion settings from transferring through from your source spreadsheet to your target spreadsheet. Overlooking this will cause you all kinds of aggravation when you see how the target spreadsheet has been re-edited with the new settings. Reconstructing the previous spreadsheet format will be equally hard and time-consuming.

After you have determined the data entry area in the target spreadsheet, you create a database by specifying field names across a row and entering data in cells as you would with standard data entry. The mechanics of entering data into a database are simple: the most critical step in creating a useful database is identifying the fields properly. Database retrieval techniques work by locating the data by field names. When you are ready to enter the data into the database, you must specify the following for each datum to be entered:

- A field name
- The column width
- Type of data entry.

In determining these steps, you first need to chose the level of detail needed for each item of information. This is the same type of critical analysis you did in the chapter regarding levels of specificity for your network schedule. Once the data has been organized in the manner you wish to retrieve it in, select the appropriate column width, and determine whether to enter the data as a label or number. If you enter the database content from a standard source document, you can increase the speed of data entry by setting up the field names in the same order as that of the corresponding data items on the source document. Remember here to plan the target database carefully before establishing field names and transferring data. This program is configured with

256 fields, which is the number of columns available in any single sort.

Field names must be input as labels, even if they are numeric labels. If that is the case, use a label prefix as previously discussed. You can use more than one row for the labels; however, the program will only read the values that appear in the bottom row as the field names. Insert blank columns between fields to prevent fields from crowding and overwriting data. When deciding on the database title, remember that all field names must be unique. Any duplication of similar names will confuse the computer and cause computation string error. The format of the database includes the title, date, labels, or numbers, and all these affect how the computer sorts and searches the data.

Sorting Databases

To sort the database, you start by designing a **D**ata **R**ange. This range must include all records to be sorted and must wide enough to include all the fields in each record. If you do not include all fields when sorting, you destroy the integrity of the database because parts of one record end up with parts of other records. Having to reconstruct your database after committing this type of entry error will cause an immediate rise in your blood pressure. Also, be sure that the data range does not include the field name row. If you make this error, the field name row is considered as one of the records to be sorted and the database may be destroyed. The data range does not necessarily have to include the entire database. If part of the database already has the database organization you want, or if you do not want to sort all the records, you can selectively sort only a portion of the database.

Using Key Sorts

After choosing the **D**ata range, specify the key for the database sort. The field with the highest precedence is assigned to the Primary key, and the field with the next-highest precedence is assigned to the Secondary key. You must set a Primary key, but the use of a Secondary key is optional. Once you have specified the range to sort, specify the sort keys(s) on which to base the reordering of the records, and indicate whether the sort order (based on the sort key) is ascending or descending, then press **E**nter to execute the command. Again, make sure you use **F**ile **S**ave before sorting the database, in case you later need to restore the database to its original order.

The simplest type of database sort is the single-key database, the Primary key method. Data entered is usually sequentially or alphabetically sorted. To arrange your database sorting capability to reorder records alphabetically, first select the **D**ata range, then select **D**ata **S**ort and fill in the fields with the data entry. Select a Primary key, then type the cell address in the column containing the Primary key field. Now enter a sort order (**A**scending or **D**escending) to tell the computer the sequence in which you want the data sorted.

You can add a record to a sorted database without having to insert a row manually to place the new record in the proper position. Simply add the new record to the bottom of the current database, expand the **D**ata range, then sort the database again by using the Primary sort key.

Double-Key Sort

A double-key sort uses both a Primary key and a Secondary key. The classic example of a double-key sort is the database sort used in your phone book. In the telephone book's yellow pages, records are sorted first according to business type (the primary key), then by business name (the secondary key). To see how the double-key method can work in your databases by first sorting with one key and then another key within the first sort order, you can add a new record to the end of the Clients' Companies database and then reorder it first by state, and then by the city within the state.

Extra Keys Sorts

The Extra keys sort option allows you to specify as many as 253 sort keys to be used in addition to the Primary and Secondary key fields. These extra keys are numbered 1 through 253 and are sequenced by the computer in this precedence: Extra key 2 is used to break ties in Extra key field 1, Extra key 3 is used to break ties in Extra key 2, and so on. You assign an Extra key the same way that you assign Primary and Secondary keys. Select Extra key from the **D**ata **S**ort menu and then enter the field (column) to be used for the extra key, followed by the sort order (**A**scending or **D**escending). Enter twice to sort the database. An example of how an Extra keys sort works would be our previous example of the Clients' Companies database. Suppose that as the database grew, several of the records were tied in their City and State fields. You could sort this database on the City and State fields, specifying Company as an Extra key to break the ties. To remove an extra sort key, assign its number to the data field being used by a higher sort key. For example, to cancel Extra keys 2, select **D**ata **S**ort Extra keys, type 2, and specify the column being used by Extra keys 1.

Data Tables

In many situations in CPM scheduling, the variables in activity durations are known quantities that can be researched from company historical data. Previous project data can provide summary reports with current project variables whose exact values are already known. The results of calculations performed by using those values contain no uncertainties. Other situations, of course, will involve variables whose exact values are unknown. Spreadsheet sorts for milestone events usually fall into the first category, and sorts for delay durations often fall into the last category.

Data tables enable the scheduler to work with variables whose values are not known. With the **D**ata **W**hat-if **T**able commands, you can create tables that show how the results of formula calculations vary as the variables used in the formulas change. This extremely helpful computer forecasting function is enhanced by another function of the **D**ata **W**hat-if **T**able commands that creates *cross-tabulation tables*. A cross-tabulation table provides summary information categorized by unique information in two fields such as the total amount of delay expected between two or more long-lead items in two or more activities.

Data Table Range

The data table range is a rectangular spreadsheet area that can be any number of columns or rows, in either a horizontal or vertical configuration, that can be placed in any empty spreadsheet location. The size of the data table range is calculated by the following two factors:

1. The range has one more row than the number of input values being evaluated, and
2. The range has one more column than the number of formulas being evaluated.

The **R**ange command affects a single cell address or groups of individual cells that make up the data table range. Some range commands control the way data in certain cells appears on-screen and the way data prints out to the hard copy sorts. For example, you can change the way numbers and formulas are displayed, or justify the margin of a block of text that spans many rows of the spreadsheet. You can further determine how data will appear in column or row format. Using the range commands, you can protect certain areas of the spreadsheet from being written over with new data so that you, or other users, do not accidentally change or erase data within the spreadsheet.

Another very useful function of the range commands is to assign a name to a single cell or range of cells. By naming a column of numbers, you can create a formula that totals the numbers within the column. You can do this operation by entering the function @SUM, then entering in parentheses the range name of the column. The typical format for a data table range follows this structure:

1. The input values to be plugged into the formulas are entered down the first column.
2. The top left cell in the data table range is empty.
3. The formulas to be evaluated are entered across the first row. Each formula must refer to the input cell.
4. After the data table is calculated, each cell in the results range contains the result obtained evaluating the formula at the top of that column, with the input value at the left of that row.

The size of the data table range depends on the number of values of each variable you want to evaluate. The cells below the formula contain the various input values for one variable. These values are used for input cell number one. The cells to the right of the formula contain the various input values for the other variable. These values are used for input cell two. Be sure that the formula refers correctly to the two input cells so that the proper input values get plugged into the correct part of the formula. After the data table is calculated, each cell in the results range contains the result of evaluating the formula with the input values in that cell's row and column.

Using What-If

FastPro CPM Scheduler is programmed to remember the relationships among the cells, and does not calculate values unless instructed to do so. Therefore, you can change the value in a cell and see what happens when your formulas are recalculated with new factors. This is an extremely powerful feature for the CPM scheduler as it allows many type of analysis. For example, you can analyze the effect of an expected delay in delivery of a long-lead item, examine the effects of changing one date or more in one cell, and the results will wash through all related cells. You can then determine what kind of workaround is needed in every analysis evaluation. Situations yet to unfold can have contingency planning already developed on the computer for variables within the project schedule. To set up and solve what-if problems, enter the necessary data, numbers, and formulas into the spreadsheet, then change various numbers along the factors path until you achieve the desired end results. The program uses numeric equivalency analysis to wash through your new data change in the individual cell to all connected cells, instantly showing you the results of total change in all cells from changing the factors in one cell.

Directories / Sort Reports

FastPro CPM Scheduler has an extensive directory of sorts, each of which are configured with different summary report sorts. Enter relative data into the appropriate sort and the program automatically integrates the data and computations throughout the other related summary sorts. Timescale computations done by the program are factored from your calendar timeline entries in the network timeline sort.

The summary report sorts include other sorts.

❑ Sort by Activities

This is the main sort of the program and the heart of a CPM network schedule. We begin the network programming by entering data into this sort. This summary report breaks out each phase and the activities within each phase, assigns each activity an identifying number for program computations, the activity's duration when started and how much duration remains, percentage of activity completed to date, early & late start dates, early & late finish dates, the amount of free float assigned to the activity as well as the amount of total float that can be dedicated to or appropriated by the total project, the responsible activity subcontractor, and the activity's budgeted cost. This is the originating data generator and needs to be filled out in detail.

The activities sort sets the precedence for the network diagram format. Under this precedence diagramming format, each activity is represented by its respective activity number. You may assign any number to any activity you wish, but they must be sequential, relative to production, for the program to have job logic. The following pages are printouts of ***FastPro's*** activities sort.

FastPro CPM Scheduler

Sort By Activities

Certified Consultants

Prepared For: **Power Engineering Corporation**
Project: **Stanford University**
File Name: cc/pe/stndford/activties
Description: Cancer Research Building
Our Invoice: **95-097-PEC**
Client Invoice: **CRB-102**

Data Date: 22 Dec 93
Run Date: 14 Jan 94

Activity Number	Activity Description	Original Duration	Remaining Duration	% Comp	Early Start	Late Start	Early Finish	Late Finish
General Conditions								
00010	Subsurface Investigation	30	0	100	20 Nov 92	24 Nov 92	10 Dec 92	12 Dec 92
00100	Instructions To Bidders	12	0	100	13 Nov 92	18 Nov 92	26 Nov 92	28 Nov 92
00200	Information Available To Bidders	2	0	100	28 Nov 92	02 Dec 92	04 Dec 92	08 Dec 92
00300	Bid Forms	1	0	100	10 Dec 92	12 Dec 92	11 Dec 92	15 Dec 92
00400	Supplements To Bid Forms	3	0	100	18 Dec 92	20 Dec 92	20 Dec 92	24 Dec 92
00500	Agreement Forms	0.5	0	100	10 Dec 92	12 Dec 92	11 Dec 92	15 Dec 92
00600	Bonds And Certificates	2	0	100	12 Dec 92	14 Dec 92	14 Dec 92	18 Dec 92
00700	General Conditions	0.5	0	100	17 Dec 92	19 Dec 92	17 Dec 92	20 Dec 92
00800	Supplementary Conditions	0.5	0	100	17 Dec 92	19 Dec 92	17 Dec 92	20 Dec 92
00850	Drawings & Schedules	1	0	100	18 Dec 92	19 Dec 92	17 Dec 92	20 Dec 92
00900	Addenda & Modifications	1	0	100	19 Dec 92	21 Dec 92	20 Dec 92	25 Dec 92
Subtotal:		**53.5**	**0**	**100%**	**29**	**30**	**10**	**17**
Phase 1:	**Specifications**							
01010	Soils: Reports & Remediations	10	0	100	20 Nov 92	24 Dec 92	10 Jan 93	15 Jan 93
01020	Allowances	1	0	100	10 Dec 92	12 Dec 92	11 Dec 92	15 Dec 92
01025	Measurement & Payment	1	0	100	17 Dec 92	19 Dec 92	17 Dec 92	20 Dec 92
01030	Alternates/Alternatives	0	0	100	17 Dec 92	19 Dec 92	17 Dec 92	20 Dec 92
01040	Coordination	0.5	0	100	17 Dec 92	19 Dec 92	18 Dec 92	20 Dec 92
01050	Field Engineering	3	0	100	19 Dec 92	20 Dec 92	20 Dec 92	24 Dec 92
01060	Regulatory Requirements	1	0	100	26 Dec 92	30 Dec 92	28 Dec 92	30 Dec 92
01070	Abbreviations & Symbols	0.5	0	100	17 Dec 92	19 Dec 92	18 Dec 92	20 Dec 92
01080	Identification Systems	1	0	100	17 Dec 92	19 Dec 92	17 Dec 92	20 Dec 92
01090	Reference Standards	1	0	100	17 Dec 92	19 Dec 92	17 Dec 92	20 Dec 92
01100	Special Project Procedures	1	0	100	10 Dec 92	12 Dec 92	11 Dec 92	15 Dec 92
01200	Project Meetings	1	0	100	10 Dec 92	12 Dec 92	11 Dec 92	15 Dec 92
01300	Submittals	1	0	100	17 Dec 92	19 Dec 92	17 Dec 92	20 Dec 92
01400	Quality Control	1	0	100	10 Dec 92	12 Dec 92	11 Dec 92	15 Dec 92
01500	Construction Facilities & Temp Controls	1	0	100	17 Dec 92	19 Dec 92	17 Dec 92	20 Dec 92
01600	Material & Equipment	30	0	100	20 Dec 92	04 Jan 93	22 Jan 93	25 Jan 93
01650	Starting Of Systems/Commissioning	1	0	100	22 Jan 93	25 Jan 93	26 Jan 93	27 Jan 93
01700	Contract Closeout	1	0	100	28 Jan 93	29 Jan 93	29 Jan 93	30 Jan 93
01800	Maintenance	10	0	100	30 Jan 93	31 Jan 93	10 Feb 93	15 Feb 93
Subtotal:		**66**	**0**	**100%**	**70**	**78**	**30**	**35**
Phase 3:	**Site Work**							
02010	Subsurface Investigation	10	0	100	31 Jan 93	31 Jan 93	11 Feb 93	16 Feb 93
02050	Demolition	3	0	100	16 Feb 93	18 Feb 93	19 Feb 93	21 Feb 93
02100	Site Preparation	30	0	100	16 Feb 93	18 Feb 93	19 Feb 93	21 Feb 93
02140	Dewatering	6	2	80	20 Feb 93	22 Feb 93	25 Feb 93	27 Feb 93
02150	Shoring & Underpinning	3	0	100	25 Feb 93	27 Feb 93	27 Feb 93	30 Feb 93
02160	Excavation Support Systems	3	0	100	25 Feb 93	27 Feb 93	27 Feb 93	30 Feb 93
02170	Cofferdams							
02200	Earthwork	7	0	100	28 Feb 93	30 Feb 93	04 Mar 93	07 Mar 93
02300	Tunneling							
02350	Piles & Caissons	5	0	100	10 Mar 93	12 Mar 93	09 Mar 93	17 Mar 93
02450	Railroad Work							
02480	Marine Work							
02500	Paving & Surfacing	3	0	100	17 Mar 93	19 Mar 93	21 Mar 93	23 Mar 93
02600	Piped Utility Materials	2	0	100	17 Mar 93	19 Mar 93	21 Mar 93	23 Mar 93
02660	Water Distribution	3	0	100	17 Mar 93	19 Mar 93	21 Mar 93	23 Mar 93
02680	Fuel Distribution	3	0	100	17 Mar 93	19 Mar 93	21 Mar 93	23 Mar 93
02700	Sewer & Drainage	3	0	100	17 Mar 93	19 Mar 93	21 Mar 93	23 Mar 93
02760	Restoration Of Underground Pipelines	3	0	100	17 Mar 93	19 Mar 93	21 Mar 93	23 Mar 93
02770	Ponds & Reservoirs	3	0	100	17 Mar 93	19 Mar 93	21 Mar 93	23 Mar 93
02780	Power & Communications	3	0	100	17 Mar 93	19 Mar 93	21 Mar 93	23 Mar 93
02800	Site Improvements							
02900	Landscaping	20	0	100	17 Mar 93	19 Mar 93	21 Mar 93	23 Mar 93

Phase 4: Concrete

03100	Concrete Formwork	11	0	100	23 Mar 93	25 Mar 93	04 Apr 93	06 Apr 93
03200	Concrete Reinforcement	3	0	100	06 Apr 93	08 Apr 93	09 Apr 93	11 Apr 93
03250	Concrete Accessories							
03300	Cast-In-Place Concrete	1	0	100	11 Apr 93	12 Apr 93	13 Apr 93	15 Apr 93
03370	Concrete Curing	15	0	100	15 Apr 93	18 Apr 93	03 Mar 93	05 Mar 93
03400	Precast Concrete							
03500	Cementitious Decks							
03600	Grout	3	0	100	05 Mar 93	08 Mar 93	09 Mar 93	11 Mar 93
03700	Concrete Restoration & Cleaning							
03800	Mass Concrete							
03850	Site Clean-up	3	0	100	05 Mar 93	09 Mar 93	09 Mar 93	15 Mar 93
Subtotal:		**36**	**0**	**100%**	**45**	**49**	**35**	**39**
Phases Totals:		**268.5**	**2**	**98.78**	**105**	**78**	**78**	**82**

Phase 5: Masonry

04100	Mortar
04150	Masonry Accessories
04200	Unit Masonry
04400	Stone
04500	Masonry Restoration & Cleaning
04550	Refractories
04600	Corrision Resistant Masonry

Phase 6: Metals

05010	Metal Materials
05030	Metal Finishes
05050	Metal Fastening
05100	Structural Metal Framing
05200	Metal Joists
05300	Metal Decking
05400	Cold-Formed Metal Framing
05500	Metal Fabrications
05580	Sheet Metal Fabrications
05700	Ornamental Metal
05800	Expansion Control
05900	Hydraulic Structures

Phase 7: Wood & Plastics

06050	Fasteners & Adhesives
06100	Rough Carpentry
06130	Heavy Timber Construction
06150	Wood-Metal Systems
06170	Prefabricated Structural Wood
06200	Finish Carpentry
06300	Wood Treatment
06400	Architectural Woodwork
06500	Prefabricated Structural Plastics
06600	Plastic Fabrications

Phase 8: Thermal & Moisture Protection

07100	Waterproofing
07150	Dampproofing
07190	Vapor & Water Infiltration Barriers
07200	Insulation
07250	Fireproofing
07300	Shingles & Roofing Tiles
07400	Preformed Roofing & Cladding/Siding
07500	Membrane Roofing
07570	Traffic Topping
07600	Flashing & Sheetmetal
07700	Roof Specialities & Accessories
07800	Skylights
07900	Joint Sealers

Phase 9:	Doors & Windows
08100	Metal Doors & Frames
08200	Wood & Plastic Doors
08250	Door Opening Assemblies
08300	Special Doors
08400	Entrances & Storefronts
08500	Metal Windows
08600	Wood & Plastic Windows
08650	Special Windows
08700	Hardware
08800	Glazing
08900	Glazed Curtain Walls

Phase 10:	Finishes
09100	Metal Support Systems
09200	Lath and Plaster
09230	Aggregate Coatings
09250	Gypsum Board
09300	Tile
09400	Terrazzo
09500	Acoustical Treatment
09540	Special Surfaces
09550	Wood Flooring
09600	Stone Flooring
09630	Unit Masonry Flooring
09650	Resilient Flooring
09680	Carpet
09700	Special Flooring
09780	Floor Treatment
09800	Special Coatings
09900	Painting
09950	Wall Coverings

Phase 11:	Specialities
10100	Chalkboards & Tackboards
10150	Compartments & Cubicles
10200	Louvers & Vents
10240	Grilles & Screens
10250	Service Wall Systems
10260	Wall & Corner Guards
10270	Access Flooring
10280	Speciality Modules
10290	Pest Control
10300	Fireplaces & Stoves
10340	Prefabricated Exterior Specialties
10350	Flagpoles
10400	Identifying Devices
10450	Pedestrian Control Devices
10500	Lockers
10520	Fire Protection Specialities
10530	Protective Covers
10550	Postal Specialities
10600	Partitions
10650	Operable Partitions
10670	Storage Shelving
10700	Exterior Sun Control Devices
10750	Telephone Specialities
10800	Toilet & Bath Specialities
10880	Scales
10900	Wardrobe & Closet Specialities

Phase 12: Equipment

11010	Maintenance Equipment
11020	Security & Vault Equipment
11030	Teller & Service Equipment
11040	Ecclesiastical Equipment
11050	Library Equipment
11060	Theater & Stage Equipment
11070	Instrumental Equipment
11080	Registration Equipment
11090	Checkroom Equipment
11100	Merchantile Equipment
11110	Laundry Equipment
11120	Vending Equipment
11130	Audio-Visual Equipment
11140	Service Station Equipment
11150	Parking Control Equipment
11160	Loading Dock Equipment
11170	Solid Waste Equipment
11190	Detention Equipment
11200	Water Supply/Treatment
11280	Hydraulic Gate Valves
11300	Fluid Waste Equipment
11400	Food Service Equipment
11450	Residential Equipment
11460	Unit Kitchens
11470	Darkroom Equipment
11480	Recreational Equipment
11500	Industrial/Process Equipment
11600	Laboratory Equipment
11650	Planetarium Equipment
11660	Observatory Equipment
11700	Medical Equipment
11780	Mortuary Equipment
11850	Navigation Equipment

Phase 13: Furnishings

12050	Fabrics
12100	Artwork
12300	Manufactured Casework
12500	Window Treatment
12600	Furniture & Accessories
12670	Rugs & Mats
12700	Multiple Seating
12800	Interior Plants & Planters

Phase 14: Special Construction

13010	Air Supported Structures
13020	Integrated Assemblies
13030	Special Purpose Rooms
13080	Sound, Vibration, Seismic
13090	Radiation Protection
13100	Nuclear Reactors
13120	Pre-Engineered Structures
13150	Pools
13160	Ice Rinks
13170	Kennels
13180	Site Constructed Incinerators
13200	Gas/Liquid Storage Tanks
13220	Filter Underdrains & Media
13230	Tank Covers
13240	Oxygenation Systems
13260	Sludge Conditioning Systems
13300	Utility Control Systems
13400	Industrial/Process Controls
13500	Recording Instrumentation
13550	Transportation Controls
13600	Solar Energy Systems
13700	Wind Energy Systems
13800	Building Automation Systems
13900	Fire Suppression Systems

❑ Sort by Events

In the sort by events, we see the different structuring of the program computation factors by event dates. The columns next to the activity's description are relative to the activity. This is further connected to the computer cell address which is represented by the activity's number. The next event columns are sequence finish events connected to the *i-j* numbers in the preceding column. The sort further shows the percentage of the activity that is complete, and the activity's free and total float. The sort is then linked to the cost tracking sort by its projected cost and actual cost for the summary audit trail.

The sort by event represents the computer's registering of an event as the exact day at which an activity is just starting or finishing. Network program logic applying to all events is that all activities leading into an event can be started at that time. An activity is always preceded by an event and followed by a sequential event. Thus, an activity always has both a starting event and finishing event. In CPM, that finishing event is the starting event of the next activity.

To build the events sort database, you start by designing a data range. This range must include all events (early start, late start, early finish, and late finish) to be sorted and wide enough to include all the fields in each record. If you do not include all field ranges when sorting, you destroy the integrity of the database because parts of one record end up with parts of other records. Be sure that the data range does not include the field name row. If this error is committed, the field name row is considered one of the records to be sorted and the database may be destroyed. The data range does not necessarily have to include the entire database. If part of the sort by events database already has the database organization you want, or if you do not want to sort all the records, you can selectively sort only a portion of the database.

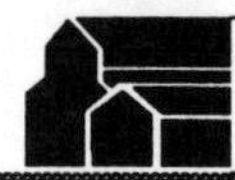

FastPro CPM Scheduler

Your Company Name Here Sort By Events

Prepared For: ***Your Client's Name Here***
Project: ***Project's Name Here***
File Name: *Your Computer File Name Here*
Description: *Project Description Here*
Our Invoice: ***Your Company's Job Number***
Client Invoice: ***Owner's Job Invoice Number***

Data Date:
Run Date:

Early Event	Late Event	Activity Description	Activity Number	I - J Number	Early Finish	Late Finish	Percent Complete
Phase 1		**General Conditions**					
0	0	Subsurface Investigation					
0	0	Instructions To Bidders					
0	0	Information Available To Bidders					
0	0	Bid Forms					
0	0	Supplements To Bid Forms					
0	0	Agreement Forms					
0	0	Bonds And Certificates					
0	0	General Conditions					
0	0	Supplementary Conditions					
0	0	Drawings & Schedules					
0	0	Addenda & Modifications					
Phase 2:		**Specifications**					
0	0	Subsurface Investigation					
0	0	Allowances					
0	0	Measurement & Payment					
0	0	Alternates/Alternatives					
0	0	Coordination					
0	0	Field Engineering					
0	0	Regulatory Requirements					
0	0	Abbreviations & Symbols					
0	0	Identification Systems					
0	0	Recordation Systems					
0	0	Reference Standards					
0	0	Special Project Procedures					
0	0	Project Meetings					
0	0	Submittals					
0	0	Quality Control					
0	0	Construction Facilities & Temp Controls					
0	0	Material & Equipment					
0	0	Starting Of Systems/Commissioning					
0	0	Contract Closeout					
0	0	Maintenance					
Phase 3:		**Commencement: Site Work**					
0	0	Subsurface Investigation					
0	0	Demolition					
0	0	Site Preparation					
0	0	Dewatering					
0	0	Shoring & Underpinning					
0	0	Excavation Support Systems					
0	0	Cofferdams					
0	0	Earthwork					
0	0	Tunneling					
0	0	Piles & Caissons					
0	0	Paving & Surfacing					
0	0	Piped Utility Materials					
0	0	Water Distribution					

Late Event	Activity Description	Activity Number	I - J Number	Early Finish	Late Finish	Percent Complete	Free Float
0	0	Fuel Distribution					
0	0	Sewer & Drainage					
0	0	Restoration Of Underground Pipelines					
0	0	Ponds & Reservoirs					
0	0	Power & Communications					
0	0	Site Improvements					
0	0	Landscaping					
Phase 4:		**Concrete**					
0	0	Concrete Formwork					
0	0	Concrete Reinforcement					
0	0	Concrete Accessories					
0	0	Cast-In-Place Concrete					
0	0	Concrete Curing					
0	0	Precast Concrete					
0	0	Cementitious Decks					
0	0	Grout					
0	0	Concrete Restoration & Cleaning					
0	0	Mass Concrete					
0	0	Site Clean-up					
Phase 5:		**Masonry**					
0	0	Mortar					
0	0	Masonry Accessories					
0	0	Unit Masonry					
0	0	Stone					
0	0	Masonry Restoration & Cleaning					
0	0	Refractories					
0	0	Corrision Resistant Masonry					
Phase 6:		**Metals**					
0	0	Metal Materials					
0	0	Metal Finishes					
0	0	Metal Fastening					
0	0	Structural Metal Framing					
0	0	Metal Joists					
0	0	Metal Decking					
0	0	Cold-Formed Metal Framing					
0	0	Metal Fabrications					
0	0	Sheet Metal Fabrications					
0	0	Ornamental Metal					
0	0	Expansion Control					
0	0	Hydraulic Structures					
Phase 7:		**Wood & Plastics**					
0	0	Fasteners & Adhesives					
0	0	Rough Carpentry					
0	0	Heavy Timber Construction					
0	0	Wood-Metal Systems					
0	0	Prefabricated Structural Wood					
0	0	Finish Carpentry					
0	0	Wood Treatment					
0	0	Architectural Woodwork					
0	0	Prefabricated Structural Plastics					
0	0	Plastic Fabrications					
Phase 8:		**Thermal & Moisture Protection**					
0	0	Waterproofing					
0	0	Dampproofing					
0	0	Vapor & Water Infiltration Barriers					
0	0	Insulation					
0	0	Fireproofing					

Late Event	Activity Description	Activity Number	I - J Number	Early Finish	Late Finish	Percent Complete	Free Float
0	0	Shingles & Roofing Tiles					
0	0	Preformed Roofing & Cladding/Siding					
0	0	Membrane Roofing					
0	0	Traffic Topping					
0	0	Flashing & Sheetmetal					
0	0	Roof Specialities & Accessories					
0	0	Skylights					
0	0	Joint Sealers					
Phase 9:		**Doors & Windows**					
0	0	Metal Doors & Frames					
0	0	Wood & Plastic Doors					
0	0	Door Opening Assemblies					
0	0	Special Doors					
0	0	Entrances & Storefronts					
0	0	Metal Windows					
0	0	Wood & Plastic Windows					
0	0	Special Windows					
0	0	Hardware					
0	0	Glazing					
0	0	Glazed Curtain Walls					
0	0	Testing					
Phase 10:		**Finishes**					
0	0	Metal Support Systems					
0	0	Lath and Plaster					
0	0	Aggregate Coatings					
0	0	Gypsum Board					
0	0	Tile					
0	0	Terrazzo					
0	0	Acoustical Treatment					
0	0	Special Surfaces					
0	0	Wood Flooring					
0	0	Stone Flooring					
0	0	Unit Masonry Flooring					
0	0	Resilient Flooring					
0	0	Carpet					
0	0	Special Flooring					
0	0	Floor Treatment					
0	0	Special Coatings					
Phase 11:		**Specialities**					
0	0	Chalkboards & Tackboards					
0	0	Compartments & Cubicles					
0	0	Louvers & Vents					
0	0	Grilles & Screens					
0	0	Service Wall Systems					
0	0	Wall & Corner Guards					
0	0	Access Flooring					
0	0	Speciality Modules					
0	0	Pest Control					
0	0	Fireplaces & Stoves					
0	0	Prefabricated Exterior Specialties					
0	0	Flagpoles					
0	0	Identifying Devices					
0	0	Pedestrian Control Devices					
0	0	Lockers					
0	0	Fire Protection Specialities					
0	0	Protective Covers					
0	0	Postal Specialities					
0	0	Partitions					

0	0	Kennels
0	0	Site Constructed Incinerators
0	0	Gas/Liquid Storage Tanks
0	0	Filter Underdrains & Media
0	0	Tank Covers
0	0	Oxygenation Systems
0	0	Sludge Conditioning Systems
0	0	Utility Control Systems
0	0	Industrial/Process Controls
0	0	Recording Instrumentation
0	0	Transportation Controls
0	0	Solar Energy Systems
0	0	Wind Energy Systems
0	0	Building Automation Systems
0	0	Fire Suppression Systems
Phase 15:		**Conveying Systems**
0	0	Dumbwaiters
0	0	Elevators
0	0	Moving Stairs & Walks
0	0	Lifts
0	0	Material Handling Systems
0	0	Hoists & Cranes
0	0	Turntables
0	0	Scaffolding
0	0	Transportation Systems
Phase 16:		**Mechanical**
0	0	Basic Materials & Methods
0	0	Mechanical Insulation
0	0	Fire Protection
0	0	Plumbing
0	0	HVAC
0	0	Heat Generation
0	0	Refrigeration
0	0	Heat Transfer
0	0	Air Handling
0	0	Air Distribution
0	0	Controls
0	0	Testing
Phase 17:		**Electrical**
0	0	Basic Materials & Methods
0	0	Power Generation
0	0	High Voltage Dist (>600v)
0	0	Service & Dist (<600v)
0	0	Lighting
0	0	Special Systems
0	0	Communications
0	0	Electric Resistance Heating
0	0	Controls
0	0	Testing
0	0	
0	0	

❑ Sort by *I-J* Numbers

In the sort by *i-j* numbers, events are logged as the exact day an activity starts or finishes. They are also dates of milestone completions. Events are assigned an identification number for computer processing. The starting event number is the *i* number and the completion eventnumber is the *j* number. The *i-j* number is used as a relative cell address for the activity's data recordation. If CPM diagram were to be prepared using random activity numbering, all activity numbers in the entire network would have to be renumbered to allow any new or changed activity to be sequential with the other activities.

Therefore it is a basic CPM requirement that, when event numbers are assigned, the finishing event number at the head of the arrow must be greater that the starting event number at the tail of the arrow, and the *j* value of each activity must be greater than its *i* value. A typical CPM network can involve hundreds of separate activities that must have flexibilities in scheduling, so the experienced project scheduler assigns the activities *i-j* numbers only *after* the entire network has been completed and is ready for its first cooking or trial-run computation.

Using the vertical and horizontal axes of graph coordinates, *i-j* events can be displayed in either plane. The vertical numbering method is more widely used, which numbers all events in a vertical column in sequence from top to bottom that equates to a parallel timeline with those groups of activities moving from left to right. There is no significance to the event numbers themselves except as a means of identifying an activity, so if the CPM format of keeping the *j* value of each activity greater than its *i* value is used, blank cells can be left in the numbering system so that spare numbers are available for changes or addition work that may come up. Sequential *i-j* numbering provides this flexibility in scheduling, while also providing the computer with program logic data for events and activities locations on the network diagram.

Your Company Name Here **Sort By I - J Numbers**

Prepared For: ***Your Client's Name Here***
Project: ***Project's Name Here*** Page: 1 of 4
File Name: *Your Computer File Name Here*
Description: *Project Description Here*
Our Invoice: ***Your Company's Job Number***
Client Invoice: ***Owner's Job Invoice Number***

I Node	J Node	Activity Description	Original Duration	Remain Duration	% Comp	Preceeding I Number	Succeeding J Number
1	3	Subsurface Investigation				0	4
1	3	Instructions To Bidders				0	4
1	3	Information Available To Bidders				0	4
1	3	Bid Forms				0	4
1	3	Supplements To Bid Forms				0	4
1	3	Agreement Forms				0	4
1	3	Bonds And Certificates				0	4
2	4	General Conditions				1	5
2	4	Supplementary Conditions				1	5
2	4	Drawings & Schedules				1	5
2	4	Addenda & Modifications				1	5
Phase 2:		**Specifications**					
3	5	Subsurface Investigation				2	6
3	5	Allowances				2	6
3	5	Measurement & Payment				2	6
3	5	Alternates/Alternatives				2	6
3	6	Coordination				2	7
3	6	Field Engineering				2	7
3	6	Regulatory Requirements				2	7
3	7	Abbreviations & Symbols				2	8
4	7	Identification Systems				3	8
4	7	Recordation Systems				3	8
4	7	Reference Standards				3	8
4	8	Special Project Procedures				3	9
4	8	Project Meetings				3	9
4	8	Submittals				3	9
4	8	Quality Control				3	9
4	9	Construction Facilities & Temp Controls				3	10
4	9	Material & Equipment				3	10
4	9	Starting Of Systems/Commissioning				3	10
4	9	Contract Closeout				3	10
4	9	Maintenance				3	10
Phase 3:		**Commencement: Site Work**					
5	10	Subsurface Investigation				4	11
6	15	Demolition				5	16
7	17	Site Preparation				6	18
8	18	Dewatering				7	19
10	20	Shoring & Underpinning				9	21
15	22	Excavation Support Systems				14	23
17	26	Cofferdams				16	27
20	28	Earthwork				19	29
23	32	Tunneling				22	33
25	38	Piles & Cassions				24	39
30	45	Railroad Work				29	46
35	48	Marine Work				34	49
40	49	Paving & Surfacing				39	50
45	50	Piped Utility Materials				44	51
50	60	Water Distribution				49	61

53	68	Fuel Distribution	52	69
55	73	Sewer & Drainage	54	74
60	75	Restoration Of Underground Pipelines	59	76
65	80	Ponds & Reservoirs	64	81
70	83	Power & Communications	69	84
75	88	Site Improvements	74	89
80	90	Landscaping	79	91
Phase 4:		**Concrete**		
90	110	Concrete Formwork	89	111
92	117	Concrete Reinforcement	91	118
94	120	Concrete Accessories	93	121
95	125	Cast-In-Place Concrete	94	126
96	128	Concrete Curing	95	129
97	130	Precast Concrete	96	131
98	132	Cementitious Decks	97	133
100	135	Grout	99	136
102	137	Concrete Restoration & Cleaning	101	138
105	138	Mass Concrete	104	139
108	139	Site Clean-up	107	140
I	J	**Masonry**		
Phase 5:				1
110	140	Mortar	109	141
111	142	Masonry Accessories	110	143
112	143	Unit Masonry	111	144
115	145	Stone	114	146
117	148	Masonry Restoration & Cleaning	116	149
118	150	Refractories	117	151
119	157	Corrision Resistant Masonry	118	158
Phase 6:		**Metals**		
120	230	Metal Materials	119	231
121	235	Metal Finishes	120	236
123	240	Metal Fastening	122	241
124	245	Structural Metal Framing	123	246
125	250	Metal Joists	124	251
126	255	Metal Decking	125	256
127	260	Cold-Formed Metal Framing	126	261
128	265	Metal Fabrications	127	266
129	270	Sheet Metal Fabrications	128	271
130	275	Ornamental Metal	129	276
133	280	Expansion Control	132	281
135	285	Hydraulic Structures	134	286
Phase 7:		**Wood & Plastics**		
140	290	Fasteners & Adhesives	139	291
142	295	Rough Carpentry	141	296
145	300	Heavy Timber Construction	144	301
147	305	Wood-Metal Systems	146	306
148	310	Prefabricated Structural Wood	147	311
150	315	Finish Carpentry	149	316
151	320	Wood Treatment	150	321
153	325	Architectural Woodwork	152	326
155	330	Prefabricated Structural Plastics	154	331
157	335	Plastic Fabrications	156	336
Phase 8:		**Thermal & Moisture Protection**		
160	340	Waterproofing	159	341
162	345	Dampproofing	161	346
163	350	Vapor & Water Infiltration Barriers	162	351
165	355	Insulation	164	356
166	360	Fireproofing	165	361
168	365	Shingles & Roofing Tiles	167	366
169	370	Preformed Roofing & Cladding/Siding	168	371

170	375	Membrane Roofing	169	376
173	380	Traffic Topping	172	381
175	385	Flashing & Sheetmetal	174	386
178	390	Roof Specialities & Accessories	177	391
180	400	Skylights	179	401
182	405	Joint Sealers	181	406
Phase 9:		**Doors & Windows**		
185	410	Metal Doors & Frames	184	411
187	412	Wood & Plastic Doors	186	413
188	415	Door Opening Assemblies	187	416
190	417	Special Doors	189	418
191	419	Entrances & Storefronts	190	420
193	420	Metal Windows	192	421
195	421	Wood & Plastic Windows	194	422
196	426	Special Windows	195	427
197	427	Hardware	196	428
198	429	Glazing	197	430
200	430	Glazed Curtain Walls	199	431
205	435	Testing	204	436
Phase 10:		**Finishes**		
210	440	Metal Support Systems	209	441
214	442	Lath and Plaster	213	443
215	445	Aggregate Coatings	214	446
217	447	Gypsum Board	216	448
220	448	Tile	219	449
223	450	Terrazzo	222	451
225	452	Acoustical Treatment	224	453
227	455	Special Surfaces	226	456
229	458	Wood Flooring	228	459
230	460	Stone Flooring	229	461
235	461	Unit Masonry Flooring	234	462
240	463	Resilient Flooring	239	464
241	465	Carpet	240	466
243	467	Special Flooring	242	468
245	470	Floor Treatment	244	471
247	473	Special Coatings	246	474
248	474	Painting	247	475
250	475	Wall Coverings	249	476
I	J	**Specialities**		
Phase 11:				1
253	480	Chalkboards & Tackboards	252	481
258	482	Compartments & Cubicles	257	483
260	484	Louvers & Vents	259	485
261	485	Grilles & Screens	260	486
265	487	Service Wall Systems	264	488
268	488	Wall & Corner Guards	267	489
270	490	Access Flooring	269	491
275	495	Speciality Modules	274	496
280	500	Pest Control	279	501
285	505	Fireplaces & Stoves	284	506
290	515	Prefabricated Exterior Specialties	289	516
295	520	Flagpoles	294	521
300	523	Identifying Devices	299	524
305	525	Pedestrian Control Devices	304	526
310	527	Lockers	309	528
315	530	Fire Protection Specialities	314	531
320	535	Protective Covers	319	536
322	537	Postal Specialities	321	538
325	540	Partitions	324	541
328	542	Operable Partitions	327	543
329	543	Storage Shelving	328	544

330	545	Exterior Sun Control Devices	329	546
335	550	Telephone Specialities	334	551
340	555	Toilet & Bath Specialities	339	556
342	560	Scales	341	561
345	565	Wardrobe & Closet Specialities	344	566
Phase 12:		**Equipment**		
350	570	Maintenance Equipment	349	571
355	572	Security & Vault Equipment	354	573
357	573	Teller & Service Equipment	356	574
360	575	Ecclesiastical Equipment	359	576
365	576	Library Equipment	364	577
370	577	Theater & Stage Equipment	369	578
375	578	Instrumental Equipment	374	579
380	579	Registration Equipment	379	580
383	580	Checkroom Equipment	382	581
385	581	Merchantile Equipment	384	582
389	582	Laundry Equipment	388	583
340	583	Vending Equipment	339	584
345	584	Audio-Visual Equipment	344	585
350	590	Service Station Equipment	349	591
355	595	Parking Control Equipment	354	596
360	560	Loading Dock Equipment	359	561
365	565	Solid Waste Equipment	364	566
370	568	Detention Equipment	369	569
373	569	Water Supply/Treatment	372	570
375	570	Hydraulic Gate Valves	374	571
380	571	Fluid Waste Equipment	379	572
385	575	Food Service Equipment	384	576
390	580	Residential Equipment	389	581
395	583	Unit Kitchens	394	584
400	585	Darkroom Equipment	399	586
405	590	Recreational Equipment	404	591
410	595	Industrial/Process Equipment	409	596
415	600	Laboratory Equipment	414	601
420	605	Planetarium Equipment	419	606
425	610	Observatory Equipment	424	611
430	615	Medical Equipment	429	616
435	620	Mortuary Equipment	434	621
440	625	Navigation Equipment	439	626
Phase 13:		**Furnishings**		
445	630	Fabrics	444	631
450	635	Artwork	449	636
455	640	Manufactured Casework	454	641
460	643	Window Treatment	459	644
465	645	Furniture & Accessories	464	646
470	647	Rugs & Mats	469	648
475	648	Multiple Seating	474	649
480	650	Interior Plants & Planters	479	651
Phase 14:		**Special Construction**		
485	700	Air Supported Structures	484	701
490	705	Integrated Assemblies	489	706
495	710	Special Purpose Rooms	494	711
500	715	Sound, Vibration, Seismic	499	716
505	720	Radiation Protection	504	721
510	723	Nuclear Reactors	509	724
515	725	Pre-Engineered Structures	514	726
520	730	Pools	519	731
525	731	Ice Rinks	524	732
530	735	Kennels	529	736
535	740	Site Constructed Incinerators	534	741
540	743	Gas/Liquid Storage Tanks	539	744

❑ Sort by Job Logic

The logical sequence of the project's construction activities, adjusted by local limitations, is factored here in the job logic sort. The activities chosen may represent relatively large segments of the project or may be limited to small steps only. For example, a concrete slab may be a single activity on a small job, but on a larger job it will be broken into the separate steps necessary to construct it, such as excavation, sub-ex preparation, erection of forms, placing of steel, placing of concrete, finishing, curing, and stripping of forms. As the separate activities are identified and defined, the sequence relationships between them must be determined. These relationships are referred to as job logic and consist of the necessary time durations and precedence, or sequential order, of typical local construction operations that are unique to your geographic area. It is a basic fundamental in CPM that each activity has a determined starting event, which may either be its own start or the finish of the preceding activity. Activity durations cannot overlap their finish events. Therefore, job logic is established to provide a sequence of operations within practical constraints.

Established job logic in the job logic sort is then used to build program logic within the computerized CPM program. By determining the job logic, activities can have their interdependencies critically examined during all phases of the schedule, before errors costing delays and money begin. In the vertical method of event node, which is more widely used, numbers of all events are in a vertical column in a sequence from top to bottom that equates to a parallel timeline. Because of the vertical configuration, activity job logic can have logic loops in those vertical groups of activities without the scheduler realizing they are there, the error then moving from left to right on the timeline with the activity group.

Study the network diagram carefully at the beginning of the development of the schedule to confirm the job logic of the structure and to critically search for logic loops. The best method to safeguard against their inclusion in your schedule is to use of sequential *i-j* numbering. The computer sort printout cannot indicate any clues to the presence of logic loops under the use of a random *i-j* numbering system, so random numbering guarantees a greater likelihood of error by allowing logical loops to remain undiscovered. Accordingly, to make ultimate use of computer program logic, the database, which in this case is the *i-j* numbers, must be sequential.

FastPro CPM Scheduler

Your Company Name Here

Sort By Job Logic

Prepared For: ***Your Client's Name Here***
Project: ***Project's Name Here***
File Name: *Your Computer File Name Here*
Description: *Project Description Here*
Our Invoice: ***Your Company's Job Number***
Client Invoice: ***Owner's Job Invoice Number***

Data Date: *Data Entry Date*
Run Date: *Printout Date*

I - J Number	Activity Number	Task or Activity Description	Phase Number	Critical Path	Early Start	Late Start	Total Float	Sequence Number:
		General Conditions						
		Subsurface Investigation	1					1
		Instructions To Bidders	1					2
		Information Available To Bidders	1					3
		Bid Forms	1					4
		Supplements To Bid Forms	1					5
		Agreement Forms	1					6
		Bonds And Certificates	1					7
		General Conditions	1					8
		Supplementary Conditions	1					9
		Drawings & Schedules	1					10
		Addenda & Modifications	1					11
		Specifications						
		Subsurface Investigation	2					12
		Allowances	2					13
		Measurement & Payment	2					14
		Alternates/Alternatives	2					15
		Coordination	2					16
		Field Engineering	2					17
		Regulatory Requirements	2					18
		Abbreviations & Symbols	2					19
		Identification Systems	2					20
		Recordation Systems	2					21
		Reference Standards	2					22
		Special Project Procedures	2					23
		Project Meetings	2					24
		Submittals	2					25
		Quality Control	2					26
		Construction Facilities & Temp Controls	2					27
		Material & Equipment	2					28
		Starting Of Systems/Commissioning	2					29
		Contract Closeout	2					30
		Maintenance	2					31
		Commencement: Site Work						
		Subsurface Investigation	3					32
		Demolition	3					33
		Site Preparation	3					34
		Dewatering	3					35
		Shoring & Underpinning	3					36
		Excavation Support Systems	3					37
		Cofferdams	3					38
		Earthwork	3					39
		Tunneling	3					40
		Piles & Cassions	3					41
		Railroad Work	3					42
		Marine Work	3					43
		Paving & Surfacing	3					44
		Piped Utility Materials	3					45
		Water Distribution	3					46
		Fuel Distribution	3					47
		Sewer & Drainage	3					48
		Restoration Of Underground Pipelines	3					49
		Ponds & Reservoirs	3					50
		Power & Communications	3					51
		Site Improvements	3					52
		Landscaping	3					53
		Concrete						
		Concrete Formwork	4					54
		Concrete Reinforcement	4					55
		Concrete Accessories	4					56
		Cast-In-Place Concrete	4					57
		Concrete Curing	4					58
		Precast Concrete	4					59
		Cementitious Decks	4					60
		Grout	4					61
		Concrete Restoration & Cleaning	4					62
		Mass Concrete	4					63

Section	Division	No.
Site Clean-up	**4**	64
Masonry		
Mortar	**5**	65
Masonry Accessories	**5**	66
Unit Masonry	**5**	67
Stone	**5**	68
Masonry Restoration & Cleaning	**5**	69
Refractories	**5**	70
Corrision Resistant Masonry	**5**	71
Metals		
Metal Materials	**6**	72
Metal Finishes	**6**	73
Metal Fastening	**6**	74
Structural Metal Framing	**6**	75
Metal Joists	**6**	76
Metal Decking	**6**	77
Cold-Formed Metal Framing	**6**	
Metal Fabrications	**6**	78
Sheet Metal Fabrications	**6**	79
Ornamental Metal	**6**	80
Expansion Control	**6**	81
Hydraulic Structures	**6**	82
Wood & Plastics		
Fasteners & Adhesives	**7**	83
Rough Carpentry	**7**	84
Heavy Timber Construction	**7**	85
Wood-Metal Systems	**7**	86
Prefabricated Structural Wood	**7**	87
Finish Carpentry	**7**	88
Wood Treatment	**7**	89
Architectural Woodwork	**7**	90
Prefabricated Structural Plastics	**7**	91
Plastic Fabrications	**7**	92
Thermal & Moisture Protection		
Waterproofing	**8**	93
Dampproofing	**8**	94
Vapor & Water Infiltration Barriers	**8**	95
Insulation	**8**	96
Fireproofing	**8**	97
Shingles & Roofing Tiles	**8**	98
Preformed Roofing & Cladding/Siding	**8**	99
Membrane Roofing	**8**	100
Traffic Topping	**8**	101
Flashing & Sheetmetal	**8**	102
Roof Specialities & Accessories	**8**	103
Skylights	**8**	104
Joint Sealers	**8**	105
Doors & Windows		
Metal Doors & Frames	**9**	106
Wood & Plastic Doors	**9**	107
Door Opening Assemblies	**9**	108
Special Doors	**9**	109
Entrances & Storefronts	**9**	110
Metal Windows	**9**	111
Wood & Plastic Windows	**9**	112
Special Windows	**9**	113
Hardware	**9**	114
Glazing	**9**	115
Glazed Curtain Walls	**9**	116
Testing	**9**	117
Finishes		
Metal Support Systems	**10**	118
Lath and Plaster	**10**	119
Aggregate Coatings	**10**	120
Gypsum Board	**10**	121
Tile	**10**	123
Terrazzo	**10**	124
Acoustical Treatment	**10**	125
Special Surfaces	**10**	126
Wood Flooring	**10**	127
Stone Flooring	**10**	128
Unit Masonry Flooring	**10**	129
Resilient Flooring	**10**	130
Carpet	**10**	131
Special Flooring	**10**	132
Floor Treatment	**10**	133
Special Coatings	**10**	134
Painting	**10**	135
Wall Coverings	**10**	136

Specialities

Specialities		
Chalkboards & Tackboards	11	137
Compartments & Cubicles	11	138
Louvers & Vents	11	139
Grilles & Screens	11	140
Service Wall Systems	11	141
Wall & Corner Guards	11	142
Access Flooring	11	143
Speciality Modules	11	144
Pest Control	11	145
Fireplaces & Stoves	11	146
Prefabricated Exterior Specialties	11	147
Flagpoles	11	148
Identifying Devices	11	149
Pedestrian Control Devices	11	150
Lockers	11	151
Fire Protection Specialities	11	152
Protective Covers	11	153
Postal Specialities	11	154
Partitions	11	155
Operable Partitions	11	156
Storage Shelving	11	157
Exterior Sun Control Devices	11	158
Telephone Specialities	11	159
Toilet & Bath Specialities	11	160
Scales	11	161
Wardrobe & Closet Specialities	11	162

Equipment

Equipment		
Maintenance Equipment	12	163
Security & Vault Equipment	12	164
Teller & Service Equipment	12	165
Ecclesiastical Equipment	12	167
Library Equipment	12	168
Theater & Stage Equipment	12	169
Instrumental Equipment	12	170
Registration Equipment	12	171
Checkroom Equipment	12	172
Merchantile Equipment	12	173
Laundry Equipment	12	174
Vending Equipment	12	175
Audio-Visual Equipment	12	176
Service Station Equipment	12	177
Parking Control Equipment	12	178
Loading Dock Equipment	12	179
Solid Waste Equipment	12	180
Detention Equipment	12	181
Water Supply/Treatment	12	182
Hydraulic Gate Valves	12	183
Fluid Waste Equipment	12	184
Food Service Equipment	12	185
Residential Equipment	12	186
Unit Kitchens	12	187
Darkroom Equipment	12	188
Recreational Equipment	12	189
Industrial/Process Equipment	12	190
Laboratory Equipment	12	191
Planetarium Equipment	12	192
Observatory Equipment	12	193
Medical Equipment	12	194
Mortuary Equipment	12	195
Navigation Equipment	12	196

Furnishings

Furnishings		
Fabrics	13	197
Artwork	13	198
Manufactured Casework	13	199
Window Treatment	13	200
Furniture & Accessories	13	201
Rugs & Mats	13	202
Multiple Seating	13	203
Interior Plants & Planters	13	204

Special Construction

Special Construction		
Air Supported Structures	14	205
Integrated Assemblies	14	206
Special Purpose Rooms	14	207
Sound, Vibration, Seismic	14	208
Radiation Protection	14	209
Nuclear Reactors	14	210
Pre-Engineered Structures	14	211
Pools	14	212
Ice Rinks	14	213
Kennels	14	214

Site Constructed Incinerators	**14**	215
Gas/Liquid Storage Tanks	**14**	216
Filter Underdrains & Media	**14**	217
Tank Covers	**14**	218
Oxygenation Systems	**14**	219
Sludge Conditioning Systems	**14**	220
Utility Control Systems	**14**	221
Industrial/Process Controls	**14**	222
Recording Instrumentation	**14**	223
Transportation Controls	**14**	224
Solar Energy Systems	**14**	225
Wind Energy Systems	**14**	226
Building Automation Systems	**14**	227
Fire Suppression Systems	**14**	228
Conveying Systems		
Dumbwaiters	**15**	229
Elevators	**15**	230
Moving Stairs & Walks	**15**	231
Lifts	**15**	232
Material Handling Systems	**15**	233
Hoists & Cranes	**15**	234
Turntables	**15**	235
Scaffolding	**15**	236
Transportation Systems	**15**	237
Mechanical		
Basic Materials & Methods	**16**	238
Mechanical Insulation	**16**	239
Fire Protection	**16**	240
Plumbing	**16**	241
HVAC	**16**	242
Heat Generation	**16**	243
Refrigeration	**16**	244
Heat Transfer	**16**	245
Air Handling	**16**	246
Air Distribution	**16**	247
Controls	**16**	248
Testing	**16**	249
Electrical		
Basic Materials & Methods	**17**	205
Power Generation	**17**	251
High Voltage Dist (>600v)	**17**	252
Service & Dist (<600v)	**17**	253
Lighting	**17**	254
Special Systems	**17**	255
Communications	**17**	
Electric Resistance Heating	**17**	
Controls	**17**	
Testing	**17**	

❑ Sort by Total Float/Late Start

Total float is shown on the sort by total float/late start as the amount of time that an activity can be delayed without delaying the late finish event for the project completion. Total float is shared with all activities. When an activity has a certain amount of total float, it can be used without tighter scheduling constraints occurring to all of the other critical path activities. Late start of the activity will reduce this availability and therefore must be computed against the total float to provide accurate computation.

Free float is not shared with other activities as is total float, so free float provides the only true measure of how many days an activity can be delayed or extended without delaying any of the other activities. It must also be added as part of the total float of preceding activities. In the sort by total float/late start, the network schedule is linked to show the CPM network with required activities time duration's in days. Free float would be a portion of each activity's noncritical duration. Total float would be the sum of each activity's remaining free float that could be committed to the critical path. Total float shown on the sort by total float/late start is computed twice by the computer, then plotted once on the spreadsheet for early start and finish events and plotted again for late start and finish events.

Line float is the amount of available slack time per line item on the computer network. In this software program, line float time is easy to compute because it is simply the difference between the early and late dates for an activity. It represents the available time between the earliest time in which an activity can be accomplished (based upon the status of the project to date) and the latest time by which it must finish for the project to finish by its event deadline.

When the critical path has been delayed and production is now behind schedule, the earliest starting event when an activity can begin is now past the latest time in which it can be completed to stay on schedule. This is known as negative float. Because the activity has no float its completion time is reduced to critical duration and, if it is behind, the difference between the early and late dates on the sort by total float/late start is less than zero. Negative float shows how far behind schedule the activity is, and if it is a critical path activity, how it shows late the project completion will be.

FastPro CPM Scheduler

Your Company Name Here | **Sort By Total Float/Late Start**

Prepared For: ***Your Client's Name Here***
Project: ***Project's Name Here*** | **Page: 1 of 3**
File Name: *Your Computer File Name Here* | Data Date: *Data Entry Date*
Description: *Project Description Here* | Run Date: *Printout Date*
Our Invoice: ***Your Company's Job Number***
Client Invoice: ***Owner's Job Invoice Number***

Phase 1

I Node	J Node	Activity Description	Duration Est	% Comp	Actual Start	Actual Finish	Free Float	Total Float
		Subsurface Investigation						
		Instructions To Bidders						
		Information Available To Bidders						
		Bid Forms						
		Supplements To Bid Forms						
		Agreement Forms						
		Bonds And Certificates						
		General Conditions						
		Supplementary Conditions						
		Drawings & Schedules						
		Addenda & Modifications						
Phase 2:		**Specifications**						
		Subsurface Investigation						
		Allowances						
		Measurement & Payment						
		Alternates/Alternatives						
		Coordination						
		Field Engineering						
		Regulatory Requirements						
		Abbreviations & Symbols						
		Identification Systems						
		Recordation Systems						
		Reference Standards						
		Special Project Procedures						
		Project Meetings						
		Submittals						
		Quality Control						
		Construction Facilities & Temp Controls						
		Material & Equipment						
		Starting Of Systems/Commissioning						
		Contract Closeout						
		Maintenance						
Phase 3:		**Commencement: Site Work**						
		Subsurface Investigation						
		Demolition						
		Site Preparation						
		Dewatering						
		Shoring & Underpinning						
		Excavation Support Systems						
		Cofferdams						
		Earthwork						
		Tunneling						
		Piles & Caissons						
		Railroad Work						
		Marine Work						
		Paving & Surfacing						
		Piped Utility Materials						
		Water Distribution						
		Fuel Distribution						
		Sewer & Drainage						
		Restoration Of Underground Pipelines						
		Ponds & Reservoirs						

Power & Communications
Site Improvements
Landscaping

Phase 4: **Concrete**

Concrete Formwork
Concrete Reinforcement
Concrete Accessories
Cast-In-Place Concrete
Concrete Curing
Precast Concrete
Cementitious Decks
Grout
Concrete Restoration & Cleaning
Mass Concrete
Site Clean-up

Phase 5: **Masonry**

Mortar
Masonry Accessories
Unit Masonry
Stone
Masonry Restoration & Cleaning
Refractories
Corrision Resistant Masonry

Phase 6: **Metals**

Metal Materials
Metal Finishes
Metal Fastening
Structural Metal Framing
Metal Joists
Metal Decking
Cold-Formed Metal Framing
Metal Fabrications
Sheet Metal Fabrications
Ornamental Metal
Expansion Control
Hydraulic Structures

Phase 7: **Wood & Plastics**

Fasteners & Adhesives
Rough Carpentry
Heavy Timber Construction
Wood-Metal Systems
Prefabricated Structural Wood
Finish Carpentry
Wood Treatment
Architectural Woodwork
Prefabricated Structural Plastics
Plastic Fabrications

Phase 8: **Thermal & Moisture Protection**

Waterproofing
Dampproofing
Vapor & Water Infiltration Barriers
Insulation
Fireproofing
Shingles & Roofing Tiles
Preformed Roofing & Cladding/Siding
Membrane Roofing
Traffic Topping
Flashing & Sheetmetal
Roof Specialities & Accessories
Skylights
Joint Sealers

Phase 9: **Doors & Windows**

Metal Doors & Frames
Wood & Plastic Doors

Door Opening Assemblies
Special Doors
Entrances & Storefronts
Metal Windows
Wood & Plastic Windows
Special Windows
Hardware
Glazing
Glazed Curtain Walls
Testing

Phase 10: Finishes

Metal Support Systems
Lath and Plaster
Aggregate Coatings
Gypsum Board
Tile
Terrazzo
Acoustical Treatment
Special Surfaces
Wood Flooring
Stone Flooring
Unit Masonry Flooring
Resilient Flooring
Carpet
Special Flooring
Floor Treatment
Special Coatings
Painting
Wall Coverings

Phase 11: Specialities

Chalkboards & Tackboards
Compartments & Cubicles
Louvers & Vents
Grilles & Screens
Service Wall Systems
Wall & Corner Guards
Access Flooring
Speciality Modules
Pest Control
Fireplaces & Stoves
Prefabricated Exterior Specialties
Flagpoles
Identifying Devices
Pedestrian Control Devices
Lockers
Fire Protection Specialities
Protective Covers
Postal Specialities
Partitions
Operable Partitions
Storage Shelving
Exterior Sun Control Devices
Telephone Specialities
Toilet & Bath Specialities
Scales
Wardrobe & Closet Specialities

Phase 12: Equipment

Maintenance Equipment
Security & Vault Equipment
Teller & Service Equipment
Ecclesiastical Equipment
Library Equipment
Theater & Stage Equipment
Instrumental Equipment
Registration Equipment

- Checkroom Equipment
- Merchantile Equipment
- Laundry Equipment
- Vending Equipment
- Audio-Visual Equipment
- Service Station Equipment
- Parking Control Equipment
- Loading Dock Equipment
- Solid Waste Equipment
- Detention Equipment
- Water Supply/Treatment
- Hydraulic Gate Valves
- Fluid Waste Equipment
- Food Service Equipment
- Residential Equipment
- Unit Kitchens
- Darkroom Equipment
- Recreational Equipment
- Industrial/Process Equipment
- Laboratory Equipment
- Planetarium Equipment
- Observatory Equipment
- Medical Equipment
- Mortuary Equipment
- Navigation Equipment

Phase 13: Furnishings

- Fabrics
- Artwork
- Manufactured Casework
- Window Treatment
- Furniture & Accessories
- Rugs & Mats
- Multiple Seating
- Interior Plants & Planters

Phase 14: Special Construction

- Air Supported Structures
- Integrated Assemblies
- Special Purpose Rooms
- Sound, Vibration, Seismic
- Radiation Protection
- Nuclear Reactors
- Pre-Engineered Structures
- Pools
- Ice Rinks
- Kennels
- Site Constructed Incinerators
- Gas/Liquid Storage Tanks
- Filter Underdrains & Media
- Tank Covers
- Oxygenation Systems
- Sludge Conditioning Systems
- Utility Control Systems
- Industrial/Process Controls
- Recording Instrumentation
- Transportation Controls
- Solar Energy Systems
- Wind Energy Systems
- Building Automation Systems
- Fire Suppression Systems

Phase 15: Conveying Systems

- Dumbwaiters
- Elevators
- Moving Stairs & Walks
- Lifts
- Material Handling Systems

❑ Cost by Activity Number

The crucial tasks of monitoring the project field service costs as separate and distinct from the architect's design costs are handled by the cost by activity number sort. The amount of the field costs is a function of the size and classifications of the on-site field labor forces assigned to the project, field office overhead costs, materials and supplies, support services from the home office, vehicle leases and fuel charges, field office share of corporate general and administrative (G&A) costs, outside consultants or contract services, and job profit. By tabulating the monthly accumulations of budget and actual field service costs, then graphically plotting each amount on this time-versus-cost chart (similar in form to the traditional construction S-curve chart), you can make a visual comparison that will clearly show not only the status of the contract at any given time, and will also show the project manager or owner any change in trend toward either a savings or a cost overrun. By also plotting a curve representing the amounts invoiced to the owner for such field services, you can provide additional dimension of usable data is provided to the owner.

To prepare and maintain the cost tracking chart on the computer program, enter the items into the cost by activity number program file. This sort will calculate the estimated, current, and projected costs for every item you enter, in every activity you're using. Accurate data for cost tracking requires regular inputs from the site supervisor on all field costs and hours of work in each classification at the project site. The project scheduler arranges to receive all such field data from the project manager, at the end of each week. Typically, the monthly pay estimates of the project's contractors' work for progress payments are submitted by the 25th of the month, and all submittals are in the office for payment before the end of the month. Billings from the architect to the owner, however, usually are based on the closing date of the end of the month. Therefore, weekly tabulation of these data should not interfere with the project manager's review of the contractors' pay requests. The cost tracking should be done on a unit basis in all activity items. These are then used to produce the audit trail.

One of the major advantages of computerized CPM is its ability to link items and activities costs to the network schedule milestones. Most major computer programs have linked cost tracking with milestones that exist within those databases to identify the purchasing-versus-cost relationships that combined provide an audit trail of those costs. By setting up the cost by activity number in your software program, you create the schedule's audit trail.

FastPro CPM Scheduler

Your Company Name Here

Cost By Activity Number

Prepared For: *Your Client's Name Here*
Project: *Project's Name Here*
File Name: *Your Computer File Name Here*
Description: *Project Description Here*
Our Invoice: *Your Company's Job Number*
Client Invoice: *Owner's Job Invoice Number*

Data Date: *Data Entry Date*
Run Date: *Printout Date*

Page: 1 of 4

Phase 1

I - J Number	Activity Description	Line Float	Avail Float	Neg Float	Estimated	Cost Current	Projected Due
	Subsurface Investigation						
	Instructions To Bidders						
	Information Available To Bidders						
	Bid Forms						
	Supplements To Bid Forms						
	Agreement Forms						
	Bonds And Certificates						
	General Conditions						
	Supplementary Conditions						
	Drawings & Schedules						
	Addenda & Modifications						
Phase 2:	**Specifications**						
	Subsurface Investigation						
	Allowances						
	Measurement & Payment						
	Alternates/Alternatives						
	Coordination						
	Field Engineering						
	Regulatory Requirements						
	Abbreviations & Symbols						
	Identification Systems						
	Recordation Systems						
	Reference Standards						
	Special Project Procedures						
	Project Meetings						
	Submittals						
	Quality Control						
	Construction Facilities & Temp Controls						
	Material & Equipment						
	Starting Of Systems/Commissioning						
	Contract Closeout						
	Maintenance						
Phase 3:	**Commencement: Site Work**						
	Subsurface Investigation						
	Demolition						
	Site Preparation						
	Dewatering						
	Shoring & Underpinning						
	Excavation Support Systems						
	Cofferdams						
	Earthwork						
	Tunneling						
	Piles & Cassions						
	Railroad Work						
	Marine Work						
	Paving & Surfacing						
	Piped Utility Materials						
	Water Distribution						
	Fuel Distribution						
	Sewer & Drainage						
	Restoration Of Underground Pipelines						
	Ponds & Reservoirs						
	Power & Communications						
	Site Improvements						
	Landscaping						
Phase 4:	**Concrete**						
	Concrete Formwork						
	Concrete Reinforcement						
	Concrete Accessories						
	Cast-In-Place Concrete						
	Concrete Curing						
	Precast Concrete						
	Cementitious Decks						
	Grout						
	Concrete Restoration & Cleaning						
	Mass Concrete						
	Site Clean-up						
Phase 5:	**Masonry**						
	Mortar						
	Masonry Accessories						
	Unit Masonry						
	Stone						
	Masonry Restoration & Cleaning						
	Refractories						
	Corrision Resistant Masonry						
Phase 6:	**Metals**						
	Metal Materials						
	Metal Finishes						
	Metal Fastening						

Structural Metal Framing
Metal Joists
Metal Decking
Cold-Formed Metal Framing
Metal Fabrications
Sheet Metal Fabrications
Ornamental Metal
Expansion Control
Hydraulic Structures

Phase 7: Wood & Plastics

Fasteners & Adhesives
Rough Carpentry
Heavy Timber Construction
Wood-Metal Systems
Prefabricated Structural Wood
Finish Carpentry
Wood Treatment
Architectural Woodwork
Prefabricated Structural Plastics
Plastic Fabrications

Phase 8: Thermal & Moisture Protection

Waterproofing
Dampproofing
Vapor & Water Infiltration Barriers
Insulation
Fireproofing
Shingles & Roofing Tiles
Preformed Roofing & Cladding/Siding
Membrane Roofing
Traffic Topping
Flashing & Sheetmetal
Roof Specialities & Accessories
Skylights
Joint Sealers

Phase 9: Doors & Windows

Metal Doors & Frames
Wood & Plastic Doors
Door Opening Assemblies
Special Doors
Entrances & Storefronts
Metal Windows
Wood & Plastic Windows
Special Windows
Hardware
Glazing
Glazed Curtain Walls
Testing

Phase 10: Finishes

Metal Support Systems
Lath and Plaster
Aggregate Coatings
Gypsum Board
Tile
Terrazzo
Acoustical Treatment
Special Surfaces
Wood Flooring
Stone Flooring
Unit Masonry Flooring
Resilient Flooring
Carpet
Special Flooring
Floor Treatment
Special Coatings
Painting
Wall Coverings

Phase 11: Specialities

Chalkboards & Tackboards
Compartments & Cubicles
Louvers & Vents
Grilles & Screens
Service Wall Systems
Wall & Corner Guards
Access Flooring
Speciality Modules
Pest Control
Fireplaces & Stoves
Prefabricated Exterior Specialties
Flagpoles
Identifying Devices
Pedestrian Control Devices
Lockers
Fire Protection Specialities
Protective Covers
Postal Specialities
Partitions
Operable Partitions
Storage Shelving
Exterior Sun Control Devices
Telephone Specialities
Toilet & Bath Specialities
Scales
Wardrobe & Closet Specialities

Phase 12: Equipment			
Maintenance Equipment			
Security & Vault Equipment			
Teller & Service Equipment			
Ecclesiastical Equipment			
Library Equipment			
Theater & Stage Equipment			
Instrumental Equipment			
Registration Equipment			
Checkroom Equipment			
Merchantile Equipment			
Laundry Equipment			
Vending Equipment			
Audio-Visual Equipment			
Service Station Equipment			
Parking Control Equipment			
Loading Dock Equipment			
Solid Waste Equipment			
Detention Equipment			
Water Supply/Treatment			
Hydraulic Gate Valves			
Fluid Waste Equipment			
Food Service Equipment			
Residential Equipment			
Unit Kitchens			
Darkroom Equipment			
Recreational Equipment			
Industrial/Process Equipment			
Laboratory Equipment			
Planetarium Equipment			
Observatory Equipment			
Medical Equipment			
Mortuary Equipment			
Navigation Equipment			
Phase 13: Furnishings			
Fabrics			
Artwork			
Manufactured Casework			
Window Treatment			
Furniture & Accessories			
Rugs & Mats			
Multiple Seating			
Interior Plants & Planters			
Phase 14: Special Construction			
Air Supported Structures			
Integrated Assemblies			
Special Purpose Rooms			
Sound, Vibration, Seismic			
Radiation Protection			
Nuclear Reactors			
Pre-Engineered Structures			
Pools			
Ice Rinks			
Kennels			
Site Constructed Incinerators			
Gas/Liquid Storage Tanks			
Filter Underdrains & Media			
Tank Covers			
Oxygenation Systems			
Sludge Conditioning Systems			
Utility Control Systems			
Industrial/Process Controls			
Recording Instrumentation			
Transportation Controls			
Solar Energy Systems			
Wind Energy Systems			
Building Automation Systems			
Fire Suppression Systems			
Phase 15: Conveying Systems			
Dumbwaiters			
Elevators			
Moving Stairs & Walks			
Lifts			
Material Handling Systems			
Hoists & Cranes			
Turntables			
Scaffolding			
Transportation Systems			
Phase 16: Mechanical			
Basic Materials & Methods			
Mechanical Insulation			
Fire Protection			
Plumbing			
HVAC			
Heat Generation			
Refrigeration			
Heat Transfer			
Air Handling			
Air Distribution			
Controls			
Testing			
Phase 17: Electrical			

❑ Schedule of Anticipated Earnings

The computer uses the schedule of anticipated earnings data to provide calculations under the What-If pulldown menu. This forecasting advantage allows the project scheduler to input other numbers into the cost and time columns for computation. The next step is to critically analyze what those changes would bring if initiated now. The subtotals of each column can either be totaled or averaged, or both. Printouts can be run at any time, to freeze a particular cost scenario that appeals to the owner.

After critical analysis of all options is complete, the program returns to the original numbers without disturbing any other data. This is especially useful in construction loan interest payment scenarios. Such financial wraparounds allow the owner to maximize the potential of the loan funds without dipping into personal lines of credit, thus keeping working capital on hand and available.

This process in cost tracking and audit trail is known as trends analysis. By entering the relative data to your activities on the schedule of anticipated earnings, the computer now links the activities network with the costs and expenses necessary to run the network. In addition, owners using CPM systems have the advantage of control over those costs before they are paid, by having the project scheduler run trends analysis forecasting. The crucial task of monitoring the project field service costs, now linked to the project schedule, separate those cost expenses from the architect's design costs. The schedule of anticipated earnings sort tracks and calculates the classifications and amounts of the on-site field labor forces assigned to the project, as well as the field office overhead costs. An entire database can be built for expenses of materials and supplies, support services from the home office, vehicle leases and fuel charges, the field office's share of corporate general and administrative (G&A) costs, outside consultants, professional legal services, professional engineering or contract services, and separate or combined cost-to-profit ratios.

Audit trail trends analysis is an important function of the schedule of anticipated earnings sort that will provide your client with one of the prime advantages of CPM, that of cost control, and the ability to foresee the commitment of resources before they are needed.

FastPro CPM Scheduler

		Your Company Name Here	**Schedule of Anticipated Earnings**				
	Prepared For:	***Your Client's Name Here***					
	Project:	***Project's Name Here***			**Page: 1 of 3**		
	File Name:	*Your Computer File Name Here*	Data Date:	*Data Entry Date*			
	Description:	*Project Description Here*	Run Date:	*Printout Date*			
	Our Invoice:	***Your Company's Job Number***					
	Client Invoice:	***Owner's Job Invoice Number***					
Phase 1 I - J Number	**Activity Number**	**Activity Description**	**Late Finish**	**Avail Float**	**Neg Float**	**Prior Event**	**Next Event**

General Conditions

Subsurface Investigation
Instructions To Bidders
Information Available To Bidders
Bid Forms
Supplements To Bid Forms
Agreement Forms
Bonds And Certificates
General Conditions
Supplementary Conditions
Drawings & Schedules
Addenda & Modifications

Phase 2: Specifications

Subsurface Investigation
Allowances
Measurement & Payment
Alternates/Alternatives
Coordination
Field Engineering
Regulatory Requirements
Abbreviations & Symbols
Identification Systems
Recordation Systems
Reference Standards
Special Project Procedures
Project Meetings
Submittals
Quality Control
Construction Facilities & Temp Controls
Material & Equipment
Starting Of Systems/Commissioning
Contract Closeout
Maintenance

Phase 3: Commencement: Site Work

Subsurface Investigation
Demolition
Site Preparation
Dewatering
Shoring & Underpinning
Excavation Support Systems
Cofferdams
Earthwork
Tunneling
Piles & Cassions
Railroad Work
Marine Work
Paving & Surfacing
Piped Utility Materials
Water Distribution
Fuel Distribution
Sewer & Drainage
Restoration Of Underground Pipelines
Ponds & Reservoirs
Power & Communications
Site Improvements
Landscaping

Phase 4: Concrete

Concrete Formwork
Concrete Reinforcement
Concrete Accessories
Cast-In-Place Concrete
Concrete Curing
Precast Concrete
Cementitious Decks

Grout
Concrete Restoration & Cleaning
Mass Concrete
Site Clean-up

Phase 5: **Masonry**

Mortar
Masonry Accessories
Unit Masonry
Stone
Masonry Restoration & Cleaning
Refractories
Corrision Resistant Masonry

Phase 6: **Metals**

Metal Materials
Metal Finishes
Metal Fastening
Structural Metal Framing
Metal Joists
Metal Decking
Cold-Formed Metal Framing
Metal Fabrications
Sheet Metal Fabrications
Ornamental Metal
Expansion Control
Hydraulic Structures

Phase 7: **Wood & Plastics**

Fasteners & Adhesives
Rough Carpentry
Heavy Timber Construction
Wood-Metal Systems
Prefabricated Structural Wood
Finish Carpentry
Wood Treatment
Architectural Woodwork
Prefabricated Structural Plastics
Plastic Fabrications

Phase 8: **Thermal & Moisture Protection**

Waterproofing
Dampproofing
Vapor & Water Infiltration Barriers
Insulation
Fireproofing
Shingles & Roofing Tiles
Preformed Roofing & Cladding/Siding
Membrane Roofing
Traffic Topping
Flashing & Sheetmetal
Roof Specialities & Accessories
Skylights
Joint Sealers

Phase 9: **Doors & Windows**

Metal Doors & Frames
Wood & Plastic Doors
Door Opening Assemblies
Special Doors
Entrances & Storefronts
Metal Windows
Wood & Plastic Windows
Special Windows
Hardware
Glazing
Glazed Curtain Walls
Testing

Phase 10: **Finishes**

Metal Support Systems
Lath and Plaster
Aggregate Coatings
Gypsum Board
Tile
Terrazzo
Acoustical Treatment
Special Surfaces
Wood Flooring
Stone Flooring
Unit Masonry Flooring

Resilient Flooring
Carpet
Special Flooring
Floor Treatment
Special Coatings
Painting
Wall Coverings

Phase 11: Specialities

Chalkboards & Tackboards
Compartments & Cubicles
Louvers & Vents
Grilles & Screens
Service Wall Systems
Wall & Corner Guards
Access Flooring
Speciality Modules
Pest Control
Fireplaces & Stoves
Prefabricated Exterior Specialties
Flagpoles
Identifying Devices
Pedestrian Control Devices
Lockers
Fire Protection Specialities
Protective Covers
Postal Specialities
Partitions
Operable Partitions
Storage Shelving
Exterior Sun Control Devices
Telephone Specialities
Toilet & Bath Specialities
Scales
Wardrobe & Closet Specialities

Phase 12: Equipment

Maintenance Equipment
Security & Vault Equipment
Teller & Service Equipment
Ecclesiastical Equipment
Library Equipment
Theater & Stage Equipment
Instrumental Equipment
Registration Equipment
Checkroom Equipment
Merchantile Equipment
Laundry Equipment
Vending Equipment
Audio-Visual Equipment
Service Station Equipment
Parking Control Equipment
Loading Dock Equipment
Solid Waste Equipment
Detention Equipment
Water Supply/Treatment
Hydraulic Gate Valves
Fluid Waste Equipment
Food Service Equipment
Residential Equipment
Unit Kitchens
Darkroom Equipment
Recreational Equipment
Industrial/Process Equipment
Laboratory Equipment
Planetarium Equipment
Observatory Equipment
Medical Equipment
Mortuary Equipment
Navigation Equipment

Phase 13: Furnishings

Fabrics
Artwork
Manufactured Casework
Window Treatment
Furniture & Accessories
Rugs & Mats
Multiple Seating
Interior Plants & Planters

❑ Sort by Early Starts

The sort by early starts is concerned with *time* use. Here in this sort, you might allocate the hours each week of events producing float, by early starts. The sort by early starts is a complex scheduling system sort that compartmentalizes by *activity function*.

Percentages of each activity's progress toward completion is the responsibility of that activity's subcontractor but the summary report of that early start progress is the scheduler's responsibility.

These more complex sorts tend to be (or become) hierarchical. They simplify delegation of each activity's responsibilities and corresponding sorts, but de-emphasize total project objectives and are slow to adapt to changes during schedule recycling. In planning the sorts of schedule activities and operations analysis, it is important to reflect the objectives of float in the project schedule.

The degree of project scheduling control versus prime contractor and subcontractor empowerment, and the overall project matrix network, will determine in significant part whether you'll be successful in reaching the project's phases milestones on time. The sort by early starts is the averaging tool the scheduler uses between time-scaled activity events and activity control, to accelerate or constrain the activity.

FastPro CPM Scheduler

Your Company Name Here

Sort By Early Starts

Prepared For: ***Your Client's Name Here***
Project: ***Project's Name Here***
File Name: *Your Computer File Name Here*
Description: *Project Description Here*
Our Invoice: ***Your Company's Job Number***
Client Invoice: ***Owner's Job Invoice Number***

Data Date: *Data Entry Date*
Run Date: *Printout Date*

Phase 1 Start: Early Event	Late Event	Activity Description	Activity Number	I - J Number	Finish: Early Event	Late Event	Critical Path	Percent Complete	Free Float	Total Float	Neg Float

General Conditions

Subsurface Investigation
Instructions To Bidders
Information Available To Bidders
Bid Forms
Supplements To Bid Forms
Agreement Forms
Bonds And Certificates
General Conditions
Supplementary Conditions
Drawings & Schedules
Addenda & Modifications

Phase 2: **Specifications**

Subsurface Investigation
Allowances
Measurement & Payment
Alternates/Alternatives
Coordination
Field Engineering
Regulatory Requirements
Abbreviations & Symbols
Identification Systems
Recordation Systems
Reference Standards
Special Project Procedures
Project Meetings
Submittals
Quality Control
Construction Facilities & Temp Controls
Material & Equipment
Starting Of Systems/Commissioning
Contract Closeout
Maintenance

Phase 3: **Commencement: Site Work**

Subsurface Investigation
Demolition
Site Preparation
Dewatering
Shoring & Underpinning
Excavation Support Systems
Cofferdams
Earthwork
Tunneling
Piles & Cassions
Railroad Work
Marine Work
Paving & Surfacing
Piped Utility Materials
Water Distribution
Fuel Distribution
Sewer & Drainage
Restoration Of Underground Pipelines
Ponds & Reservoirs
Power & Communications
Site Improvements
Landscaping

Phase 4: **Concrete**

Concrete Formwork
Concrete Reinforcement
Concrete Accessories
Cast-In-Place Concrete
Concrete Curing
Precast Concrete
Cementitious Decks
Grout
Concrete Restoration & Cleaning
Mass Concrete
Site Clean-up

Phase 5: **Masonry**

Mortar
Masonry Accessories
Unit Masonry
Stone
Masonry Restoration & Cleaning
Refractories
Corrision Resistant Masonry

Phase 6: **Metals**

Metal Materials
Metal Finishes
Metal Fastening
Structural Metal Framing
Metal Joists
Metal Decking
Cold-Formed Metal Framing
Metal Fabrications
Sheet Metal Fabrications
Ornamental Metal
Expansion Control
Hydraulic Structures

Phase 7: **Wood & Plastics**

Fasteners & Adhesives
Rough Carpentry
Heavy Timber Construction
Wood-Metal Systems
Prefabricated Structural Wood
Finish Carpentry
Wood Treatment
Architectural Woodwork
Prefabricated Structural Plastics
Plastic Fabrications

Phase 8: **Thermal & Moisture Protection**

Waterproofing
Dampproofing
Vapor & Water Infiltration Barriers
Insulation
Fireproofing
Shingles & Roofing Tiles
Preformed Roofing & Cladding/Siding
Membrane Roofing
Traffic Topping
Flashing & Sheetmetal
Roof Specialities & Accessories
Skylights
Joint Sealers

Phase 9: **Doors & Windows**

Metal Doors & Frames
Wood & Plastic Doors
Door Opening Assemblies
Special Doors
Entrances & Storefronts
Metal Windows
Wood & Plastic Windows
Special Windows
Hardware
Glazing
Glazed Curtain Walls
Testing

Phase 10: **Finishes**

Metal Support Systems
Lath and Plaster
Aggregate Coatings
Gypsum Board
Tile
Terrazzo
Acoustical Treatment
Special Surfaces
Wood Flooring
Stone Flooring
Unit Masonry Flooring
Resilient Flooring
Carpet
Special Flooring
Floor Treatment
Special Coatings
Painting
Wall Coverings

Phase 11: **Specialities**

Chalkboards & Tackboards
Compartments & Cubicles
Louvers & Vents
Grilles & Screens
Service Wall Systems
Wall & Corner Guards
Access Flooring
Speciality Modules
Pest Control
Fireplaces & Stoves
Prefabricated Exterior Specialties
Flagpoles
Identifying Devices
Pedestrian Control Devices
Lockers
Fire Protection Specialities
Protective Covers
Postal Specialities

Partitions
Operable Partitions
Storage Shelving
Exterior Sun Control Devices
Telephone Specialities
Toilet & Bath Specialities
Scales
Wardrobe & Closet Specialities

Phase 12: Equipment

Maintenance Equipment
Security & Vault Equipment
Teller & Service Equipment
Ecclesiastical Equipment
Library Equipment
Theater & Stage Equipment
Instrumental Equipment
Registration Equipment
Checkroom Equipment
Merchantile Equipment
Laundry Equipment
Vending Equipment
Audio-Visual Equipment
Service Station Equipment
Parking Control Equipment
Loading Dock Equipment
Solid Waste Equipment
Detention Equipment
Water Supply/Treatment
Hydraulic Gate Valves
Fluid Waste Equipment
Food Service Equipment
Residential Equipment
Unit Kitchens
Darkroom Equipment
Recreational Equipment
Industrial/Process Equipment
Laboratory Equipment
Planetarium Equipment
Observatory Equipment
Medical Equipment
Mortuary Equipment
Navigation Equipment

Phase 13: Furnishings

Fabrics
Artwork
Manufactured Casework
Window Treatment
Furniture & Accessories
Rugs & Mats
Multiple Seating
Interior Plants & Planters

Phase 14: Special Construction

Air Supported Structures
Integrated Assemblies
Special Purpose Rooms
Sound, Vibration, Seismic
Radiation Protection
Nuclear Reactors
Pre-Engineered Structures
Pools
Ice Rinks
Kennels
Site Constructed Incinerators
Gas/Liquid Storage Tanks
Filter Underdrains & Media
Tank Covers
Oxygenation Systems
Sludge Conditioning Systems
Utility Control Systems
Industrial/Process Controls
Recording Instrumentation
Transportation Controls
Solar Energy Systems
Wind Energy Systems
Building Automation Systems
Fire Suppression Systems

Phase 15: Conveying Systems

Dumbwaiters
Elevators
Moving Stairs & Walks
Lifts
Material Handling Systems
Hoists & Cranes
Turntables
Scaffolding
Transportation Systems

Phase 16: Mechanical

Basic Materials & Methods

❑ Daily Field Reports

The daily field report is the fundamental sort that records actual job progress, together with all conditions that affect the work. It provides the progress reporting basis for the actual results of the schedule, as compared to projected progress. It begins the organization's standardization of reporting at the source of causes and effects- - the point of production. If actually prepared daily, it is generally considered to be the best source of job information. This is because it's are supposed to be prepared immediately as the information is being generated, with no appreciable time lapse. The inclusions, descriptions, and so on are fresh in everyone's mind. This facilitates proper dealing with all details of the work. In addition, it is prepared by those with authority and responsibility for the work. These people have an interest in the accuracy of the information, and have a significant incentive to maintain the report's accuracy and completeness. These are the people who actually witnessed the work and are recognized to be the most qualified to describe it.

Because they will be the most detailed, accurate, and complete records of all jobsite events, the daily field reports will become the cornerstones to support the actual facts. They are so valuable because of the wealth of information recorded in summary form. This data includes:

- Work and activity descriptions, separated as to the physical location and extent.
- Labor force, broken down by subcontractor, locations, and major activities.
- Equipment used and stored by the various activity subcontractors.
- Site administrative staff and facilities.
- Weather conditions and temperatures at key times of each work day.
- Change order work accomplished, with relevant details.
- Photographs taken during the day.
- List of visitors to the jobsite.
- Meetings, discussions, commitments, and conversations.

You will note that the daily field report in ***FastPro*** is two pages in length but has been condensed into one page for a demonstration example in the book. In the software program, however, the field report will expand into however many pages as you wish to build. Never use the back of a field report to continue information, it will be overlooked when faxing, during a review, or when photocopying. Because it is impossible to tell in advance what will become the important information, all categories must be accurately maintained. It is up to the central office project administration, whatever the exact authority structure within your organization, to regularly police the reports.

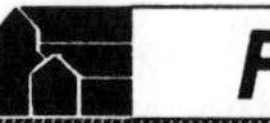

FastPro CPM Scheduler

Your Company Name Here — **Daily Field Report**

Prepared For:	***Your Client's Name Here***		
Project:	***Project's Name Here***		**Page: 1 of**
File Name:	*Your Computer File Name Here*	Data Date:	*Date of Report Input*
Description:	*Project Description Here*	Run Date:	*Printout Date of Current Copy*
Our Invoice:	***Your Company's Job Number***		
Client Invoice:	***Owner's Job Invoice Number***		

Date:	*Today's date*	**Temp**	*Begin/End*	**Critical or Principal Activity Completed:**
Phase #:	*Phase project is in*	**Wind**	*Dir/mph*	
Filed by:	*Site Supervisor preparing field report*	**Sky** / **Precip**	*Clr/Clds* / *% Rain*	
Visitors:	*Visitors to the site today (other than regular workers)*			
Reason for Visit:	*Why are they here?*			
Work Force:	*Who were the labor force present today?*			
classifications:	*What were their project's employment classification?*			
Activity performed:	*What was done by each of the above?*			
Equip on-site:	*What equipment is on-site today?*			
Equip work performed:	*What was done by each of the above?*			
Items received:	*Items received at jobsite today*			
Items sent:	*Items sent from jobsite today*			
Location of work performed:	*Where exactly in the project this work was done*			
C.O. number:	*Change order numbers if any*			
Problems:	*Problems or delays encountered today by Site Supervisor*			
Comments:	*Solutions proposed by Site Supervisor*			
Copy to:	*Project Manager* *Project Scheduler* *Owner*			

❑ Bar Chart by Early Start

By making a separate bar chart from the master bar chart by early start in ***FastPro***, for any day or weekly planner you may need, such as tracking contract documents and then making a related calendar entry in a the master network timeline sort, then saving both to a separate file, you create syncronicity between the two sorts in each distinct and dedicated file. The advantage of using CPM scheduling systems on capital projects versus bar charts is in the increased control over the interrelationships between events and higher accuracy of detailed summary sorts. The bar chart by early start provides instantaneous critical data regarding actual progress by early schedule start by activity. This data is made readily available on the computer monitor screen at any given time, and is essential for any modern production under a deadline. The CPM system with an integrated bar chart by early start can handle the hundreds of work activities on large commercial and industrial projects with ease.

Computerized CPM bar chart by early start sorts produce a bar chart printout sort, which is sorted from the activities' early starts. Bar charts are included in the weekly meetings as progress reports of scheduled progress versus actual progress. These bar chart sorts are essential tools for presentation of network scheduling details in the CPM project schedule to the owner and the project team.

Bar chart by early start charts are effective at activity scheduling and tracking as a job logic diagram. Bar charts, however, do not provide the interdependency relationship between activities, whereas CPM networking from bar chart schedules shows the dependence (constraints) between one starting activity and the finish of the activity preceding it. Further, bar charts to not allow for variable float control at those activities' events.

There are also limits to the number of activities, usually around fifty, that can be tracked on a bar chart before the chart becomes a victim of data overload and the milestones within the bar chart schedule miss their marks. Presentation bar charts are typically structured with a three-month timeline.

Many subactivities and work items within subtasks within the larger activities should be broken out with individual computer numbers for dedicated cell address data storage. All activities are made up of smaller activities that need attention and control.

FastPro CPM Scheduler

Your Company Name Here

Bar Chart by Early Start

Prepared For: ***Your Client's Name Here***
Project: ***Project's Name Here***
File Name: *Your Computer File Name Here*
Description: *Project Description Here*
Our Invoice: ***Your Company's Job Number***
Client Invoice: ***Owner's Job Invoice Number***

Data Date: *Data Entry Date*
Run Date: *Printout Date*

Activity Number	Activity Description	JAN				FEB				MAR					APR				MAY				JUN					JUL			
		6	13	20	27	3	10	17	24	3	10	17	24	31	7	14	21	28	5	12	19	26	2	9	16	23	30	7	14	21	28
00010	Subsurface Investigation																														
00100	Instructions To Bidders																														
00200	Information Available To Bidders																														
00300	Bid Forms																														
00400	Supplements To Bid Forms																														
00500	Agreement Forms																														
00600	Bonds And Certificates																														
00700	General Conditions																														
00800	Supplementary Conditions																														
00850	Drawings & Schedules																														
00900	Addenda & Modifications																														
01010	Soils: Reports & Remediations																														
01020	Allowances																														
01025	Measurement & Payment																														
01030	Alternates/Alternatives																														
01040	Coordination																														
01050	Field Engineering																														
01060	Regulatory Requirements																														
01070	Abbreviations & Symbols																														
01080	Identification Systems																														
01090	Reference Standards																														
01100	Special Project Procedures																														
01200	Project Meetings																														
01300	Submittals																														
01400	Quality Control																														
01500	Construction Facilities & Temp Controls																														
01600	Material & Equipment																														
01650	Starting Of Systems/Commissioning																														
01700	Contract Closeout																														
01800	Maintenance																														
02010	Subsurface Investigation																														
02050	Demolition																														
02100	Site Preparation																														
02140	Dewatering																														
02150	Shoring & Underpinning																														

02160	Excavation Support Systems
02170	Cofferdams
02200	Earthwork
02300	Tunneling
02350	Piles & Cassions
02450	Railroad Work
02480	Marine Work
02500	Paving & Surfacing
02600	Piped Utility Materials
02660	Water Distribution
02680	Fuel Distribution
02700	Sewer & Drainage
02760	Restoration Of Underground Pipelines
02770	Ponds & Reservoirs
02780	Power & Communications
02800	Site Improvements
02900	Landscaping
03100	Concrete Formwork
03200	Concrete Reinforcement
03250	Concrete Accessories
03300	Cast-In-Place Concrete
03370	Concrete Curing
03400	Precast Concrete
03500	Cementitious Decks
03600	Grout
03700	Concrete Restoration & Cleaning
03800	Mass Concrete
03850	Site Clean-up
04100	Mortar
04150	Masonry Accessories
04200	Unit Masonry
04400	Stone
04500	Masonry Restoration & Cleaning
04550	Refractories
04600	Corrision Resistant Masonry
05010	Metal Materials
05030	Metal Finishes
05050	Metal Fastening
05100	Structural Metal Framing
05200	Metal Joists
05300	Metal Decking
05400	Cold-Formed Metal Framing
05500	Metal Fabrications
05580	Sheet Metal Fabrications
05700	Ornamental Metal
05800	Expansion Control
05900	Hydraulic Structures
06050	Fasteners & Adhesives

06100	Rough Carpentry																														
06130	Heavy Timber Construction																														
06150	Wood-Metal Systems																														
06170	Prefabricated Structural Wood																														
06200	Finish Carpentry																														
06300	Wood Treatment																														
06400	Architectural Woodwork																														
06500	Prefabricated Structural Plastics																														
06550	Plastic Fabrications																														
07100	Waterproofing																														
07150	Dampproofing																														
07190	Vapor & Water Infiltration Barriers																														
07200	Insulation																														
07250	Fireproofing																														
07300	Shingles & Roofing Tiles																														
07400	Preformed Roofing & Cladding/Siding																														
07500	Membrane Roofing																														
07570	Traffic Topping																														
07600	Flashing & Sheetmetal																														
07700	Roof Specialities & Accessories																														
07800	Skylights																														
07900	Joint Sealers																														
08100	Metal Doors & Frames																														
08200	Wood & Plastic Doors																														
08250	Door Opening Assemblies																														
08300	Special Doors																														
08400	Entrances & Storefronts																														
08500	Metal Windows																														
08600	Wood & Plastic Windows																														
08650	Special Windows																														
08700	Hardware																														
08800	Glazing																														
08900	Glazed Curtain Walls																														
09100	Metal Support Systems																														
09200	Lath and Plaster																														
09230	Aggregate Coatings																														
09250	Gypsum Board																														
09300	Tile																														
09400	Terrazzo																														
09500	Acoustical Treatment																														
09540	Special Surfaces																														
09550	Wood Flooring																														
09600	Stone Flooring																														
09630	Unit Masonry Flooring																														
09650	Resilient Flooring																														
09680	Carpet																														
09700	Special Flooring																														
09780	Floor Treatment																														

09800	Special Coatings																														
09900	Painting																														
09950	Wall Coverings																														
10100	Chalkboards & Tackboards																														
10150	Compartments & Cubicles																														
10200	Louvers & Vents																														
10240	Grilles & Screens																														
10250	Service Wall Systems																														
10260	Wall & Corner Guards																														
10270	Access Flooring																														
10280	Speciality Modules																														
10290	Pest Control																														
10300	Fireplaces & Stoves																														
10340	Prefabricated Exterior Specialties																														
10350	Flagpoles																														
10400	Identifying Devices																														
10450	Pedestrian Control Devices																														
10500	Lockers																														
10520	Fire Protection Specialities																														
10530	Protective Covers																														
10550	Postal Specialities																														
10600	Partitions																														
10650	Operable Partitions																														
10670	Storage Shelving																														
10700	Exterior Sun Control Devices																														
10750	Telephone Specialities																														
10800	Toilet & Bath Specialities																														
10880	Scales																														
10900	Wardrobe & Closet Specialities																														
11010	Maintenance Equipment																														
11020	Security & Vault Equipment																														
11030	Teller & Service Equipment																														
11040	Ecclesiastical Equipment																														
11050	Library Equipment																														
11060	Theater & Stage Equipment																														
11070	Instrumental Equipment																														
11080	Registration Equipment																														
11090	Checkroom Equipment																														
11100	Merchantile Equipment																														
11110	Laundry Equipment																														
11120	Vending Equipment																														
11130	Audio-Visual Equipment																														
11140	Service Station Equipment																														
11150	Parking Control Equipment																														
11160	Loading Dock Equipment																														
11170	Solid Waste Equipment																														
11190	Detention Equipment																														
11200	Water Supply/Treatment																														

❑ Network Timeline

This is the timescale directory that establishes the data used by the computer for factoring time computations throughout the integrated sorts. This report anchors each phase, and the activities within each phase, to the network timeline. The master format is set up in a one-year duration but can be extended by the end user when needed.

The weekly milestones are set on Fridays to coincide with the weeky field reports meeting, and are readjustable each year by simple input of new dates. To input new calendar dates or insert new milestones, move to the relative cell address and input the data as per instructions in data entry.

The network timeline is an extremely functional sort in linkage with bar chart sorts. By making a separate bar chart from the master in ***FastPro***, for any day or weekly planner you may need, such as tracking contract documents and then making a related calendar entry in the master network timeline sort, then saving both to a separate file, you create syncronicity between the two sorts in that distinct and dedicated file.

FastPro CPM Scheduler

Your Company Name Here

Network Velocity Timeline

Prepared For: *Your Client's Name Here*
Project: *Project's Name Here*
File Name: *Your Computer File Name Here*
Description: *Project Description Here*
Our Invoice: *Your Company's Job Number*
Client Invoice: *Owner's Job Invoice Number*

Data Date: *Data Entry Date*
Run Date: *Printout Date*

Activity	Activity			JAN				FEB					MAR				APR				MAY					JUN				JUL	
Number	**Description**	6	13	20	27	3	**10**	17	24	3	10	17	24	31	7	14	21	28	5	12	19	26	2	9	16	23	30	7	14	21	28

❑ Shop Drawings Log

The shop drawings log should be started prior to commencement, so that, when the shop drawings start arriving from subcontractors and suppliers, each drawing can be properly logged into the computer for tracking. A log of shop drawing activity will allow a project scheduler to keep track of what drawings have been received and will also show where the drawings have been sent and how long they have been there. The project scheduler needs to prod each subcontractor and supplier to submit their drawings promptly. At the first job meeting, the major subcontractors and/or material suppliers should be requested to submit a preliminary shop drawing then submission schedule.

The log should include the major pieces of equipment for which shop drawings are required, and the anticipated date when each drawing will be submitted. The subcontractor and/or supplier should also include the approximate delivery date of equipment after the approved shop drawings have been returned to them. Once shop drawings have been received, the next hurdle is getting them approved in a timely manner.

The project manager and the owner should review the incoming shop drawings before the project scheduler logs them into the computer, to determine whether they conform to the project specifications and requirements. If compliance is questionable, they should contact the party who made the submission and ask them to re-examine the shop drawings for contract specifications. If there are deviations, it might be best to note them before submitting the drawings to the architect. At this point the project manager must establish credibility with the architect and engineers and must show that the shop drawings are being reviewed for compliance with the plans and specifications, and are not merely being passed through without any scrutiny whatsoever.

FastPro CPM Scheduler

Your Company Name Here

Shop Drawings Log

Prepared For: ***Your Client's Name Here***
Project: ***Project's Name Here***
File Name: *Your Computer File Name Here*
Description: *Project Description Here*
Our Invoice: ***Your Company's Job Number***
Client Invoice: ***Owner's Job Invoice Number***

Data Date: *Data Entry Date*
Run Date: *Printout Date*

Page: 1 of

Vendor Name	Vendor Address	Phone Number	Contact Person	Date Sent	Date Due	Check By:	Appvd By:	Activity Description I-J Number	Phase

❑ Submittal Items Tracking

The submittal items tracking sort is used on all long-lead and fabricated items procurement tracking. In addition, a separate file should be kept for each trade that will be submitting large numbers of drawings, such as structural steel, plumbing, HVAC, sprinkler system, and electrical.

Several software programs on the market today allow the project scheduler to create a fax modem transmittal forwarding the shop drawing to the architect or engineer, and transferring this information automatically onto a shop drawing log. The information inserted at the time of preparation is also stored and transferred to the shop drawing log. Although the shop drawing log contains material from various subcontractors, separate computer files and summary report sorts can be created for individual trades.

Care must be taken to discern which subcontractors should receive informational copies of shop drawings from the submittal items tracking sort. For instance, when a mechanical subcontractor is being sent an approved copy of their boiler shop drawing, the electrical contractor should have an informational copy to confirm the line voltage requirements. All too often a piece of equipment is ordered with electrical characteristics at variance with the voltage requirements shown in the drawings. If an error such as this can be caught in the shop drawing stage, little or no additional cost may be required to make the equipment compatible with the building's electrical system.

Your Company Name Here

Submittal Items Tracking

Prepared For: ***Your Client's Name Here***
Project: ***Project's Name Here***
File Name: *Your Computer File Name Here*
Description: *Project Description Here*
Our Invoice: ***Your Company's Job Number***
Client Invoice: ***Owner's Job Invoice Number***

Page: 1 of

Data Date: *Data Entry Date*
Run Date: *Printout Date*

Item Number	Item Name	Vendor Name & Address	Phone Number	Req'd Start	Req'd Finish	Date Returned	Date Sent	Days Over	Item Status	Description Value

❑ Correspondence Transmittals

Much of the actual day-to-day work of a project involves correspondence of schedules, unforeseen problems and changes. All correspondence relative to your network schedule must be tracked for quick recordation access.

This sort has three levels. The first records each correspondence, the date and time it was sent, its description, from whom, and to whom. It then records when the correspondence's detail is required to be on the project, when it is required to be done, the date the detail was sent ot the party concerned, and the amount of days the return is late.

The sort, through its third factoring, then track's the item's value as well as it status for an audit trail. The correspondence from vendors, materials suppliers, and various subcontractors through this sort including change orders, shop drawings, or other documents can be automatically transmitted to the concerned party and then transferred automatically to a summary report for the owner and project team members.

Your Company Name Here

Correspondence Transmittals

Prepared For: *Your Client's Name Here*
Project: *Project's Name Here*
File Name: *Your Computer File Name Here*
Description: *Project Description Here*
Our Invoice: *Your Company's Job Number*
Client Invoice: *Owner's Job Invoice Number*

Page: 1 of

Data Date: *Data Entry Date*
Run Date: *Printout Date*

Sent Date	Sent Time	Correspondence Description	To:	From:	Req'd Start	Req'd Finish	Date Returned	Date Sent	Days Over	Item Description Status	Item Description Value

❑ Change Order Tracking

Change order tracking is done by interlocking the activity number, type, and description with the status and value. This accounts for where the money is going and who's responsible for authorization and procurement. This sort also has three levels of factoring.

The first records each change order by contract number, the date and time it was authorized, authorized by whom, who will procure it, its description, from whom, and to whom. It then records when the change order must be initiated, when is must be completed, and the date the change order was sent to the party concerned.

The sort, through its third factoring, then tracks the change order's value as well as it status for audit trail. As with the correspondence transmittals sort, the change orders from activity subcontractors through the change order tracking sort can forward documentation to the concerned party and then transfer this information automatically to a summary report for the owner and project team members.

Change Order Tracking

Prepared For: ***Your Client's Name Here***
Project: ***Project's Name Here***
File Name: *Your Computer File Name Here*
Description: *Project Description Here*
Our Invoice: ***Your Company's Job Number***
Client Invoice: ***Owner's Job Invoice Number***

Data Date: *Data Entry Date*
Run Date: *Printout Date*

Page: 1 of

Activity Number	Activity Title	To:	From:	Change Number	Contract Number	Resp Contr	Supplier Vendor	Date Req'd	Date Sent	Appv'd By:	Order Status	Description Value

❑ Audit Trail Tracking

The audit trail tracking sort factors its summary report by the computer database computed from the activity's assigned *i-j* number, the activity description, the item number, the date it was linked to procurement, who sent it and to whom, and who's responsible for authorization and procurement. Tracking is done by interlocking the item number, type, and description with the status and value.

This accounts for where the money is going; by entering the appropriate data on the cost by activity number sort you can have the computer link the activities network with the costs and expenses necessary to run the network.

In addition, owners using CPM systems have the advantage of control over those costs before they are paid, by having the project scheduler run trends analysis forecasting. The crucial task of monitoring the project field service costs is now linked to the project schedule and separates those cost expenses from the architect's design costs. The audit trail tracking sort tracks and calculates these in series.

Audit trail trends analysis is an important function of your schedule estimating that will provide your client with one of the prime advantages of CPM, that of cost control and the ability to foresee the commitment of resources before they are needed. The sort can also forward documentation to the concerned party and then transfer this information automatically to a summary report for the owner and project team members.

FastPro CPM Scheduler

Your Company Name Here

Prepared For ***Your Client's Name Here***
Project: ***Project's Name Here***
File Name: *Your Computer File Name Here*
Description: *Project Description Here*
Our Invoice: ***Your Company's Job Number***
Client Invoice ***Owner's Job Invoice Number***

Audit Trail Tracking

Page: 1 of

Data Date:
Run Date:

Date Linked	Item Number	Item Description	To:	From:	I-J Number	Date Appv'd	Date Req'd	Date Sent	Check Number	Order Status	Description Value

❑ Contract/P.O. Summary

This sort provides a summary report by the computer database computed from the purchase order's assigned number, the activity description, the date it was issued, who's responsible for authorization, who sent it, and to whom. Tracking is done by original date, the revised date, and the status.

Cost changes are shown as either approved or pending. This pending column is further defined to show available resources. This allows the owner to commit resources as priorities may dictate.

FastPro CPM Scheduler

Your Company Name Here

Contract/P.O. Summary

Prepared For: ***Your Client's Name Here***
Project: ***Project's Name Here***
File Name: *Your Computer File Name Here*
Description: *Project Description Here*
Our Invoice: ***Your Company's Job Number***
Client Invoice: ***Owner's Job Invoice Number***

Page: 1 of
Data Date: *Data Entry Date*
Run Date: *Printout Date*

To:	From:	Item Description	Date Opened	Date Closed	P.O. Number	Change Issued	Appv'd Type	Status	Cost Changes Approved	Cost Changes Pending

Saving Your Sorts

When you have created a sort on the master format, the data and information will have to be saved on either your hard disk or a floppy disk. When a sort is keyed in for the first time and is displayed on the screen, it is in RAM and is only temporary. If you turn the computer off or if the power goes off, you will lose the data and information and have to re-enter it. Only when you save a sort to a disk is it saved permanently. Every time you load ***FastPro*** on your spreadsheet program, you will be able bring a saved sort back to the screen.

To save a sort you must execute a save command. ***FastPro*** provides four methods for doing this. They are:

1. Using the toolbar
2. Using the menu bar with the mouse
3. Using the menu bar with the keyboard
4. Using speed commands

FastPro contains two commands that are used to save sorts. One is the Save As command and the other is the Save command. When you begin creating a sort it does not yet have a file name. You assign it a name when you save the file. The first time you save the file you may choose either the Save or Save As command. ***FastPro*** gives you the opportunity to assign the file a name of your choosing. After you first save a sort, however, you should use the Save command to continue saving changes. ***FastPro*** automatically uses the same file name when you save it by using the Save command. The Save As command is used to save changes to an existing file under a different file name.

Speed commands are executed by pressing either one or two keys on the keyboard in conjunction with a function key, or by pressing a function key alone. In some cases, there is more than one speed command for the same function. For example, to save a sort file for the first time you can press either Shift + F12 or Alt + Shift + F2. Either method executes the Save command.

To save a file using the toolbar, complete the following steps:

1. Click on the Save icon.
2. At the Save As dialog box, key in the name of the sort.
3. Click on OK or press Enter.
4. At the Summary Info dialog box, click on OK or press Enter.

To save a sort file using the menu bar with the mouse, complete the following steps:

1. Click on File, then Save.
2. At the Save As dialog box, key in the name of the file.
3. Click on OK or press Enter.
4. At the Summary Info dialog box, click on OK or press Enter.

To save a sort file using the menu bar with the keyboard, complete the following steps:

1. Press the Alt key, key in the letter **F** for File, then the letter **S** for Save.
2. At the Save As dialog box, key in the name of the file.
3. Click on OK or press Enter.
4. At the Summary Info dialog box, click on OK or press Enter.

Cancelling a Command

Pressing the Alt key moves the cursor to the menu bar at the top of the screen. Pressing the Alt key again, or the Esc key, moves the cursor back to the sort screen. You can also move the mouse cursor to the sort screen and click the left button to move the cursor back to the sort screen. If a menu is displayed, position the mouse cursor anywhere on the sort screen outside of the menu and click the left button to remove the menu without executing a command. If you are not using a mouse, press the Alt or Esc key to remove the menu without executing a command.

To remove a dialog box from the screen without executing a command, press the Esc key. This removes the dialog box and returns the cursor to the sort screen. To close a dialog box with the mouse without executing a command, click on Cancel.

Closing a Sort

Once a sort has been saved you may want to remove it from the screen. Removing a sort from the screen does not delete the sort from the disk. To close a sort and remove it from the screen, choose File, then Close, or press Ctrl + F4. You can also close a sort by positioning the mouse cursor on the window close button (the lower button in the upper left corner of the screen), clicking the left button, and choosing Close in the menu that is displayed. This will close the sort and remove it from the screen.

Opening a Sort

When you close a sort, ***FastPro*** displays a clear screen. If you want to key in a new sort, you need to display a sort screen. To do this, complete the following steps:

1. Choose File, then New.
2. At the new dialog box, choose OK (or press Enter).

To open a previously saved sort, complete the following steps at a clear screen:

1. Choose File, then Open, or press Ctrl + F12.
2. Move the mouse cursor to the sort to be opened and click the left button. If you are not using a mouse, press the down arrow key to move the cursor to the list of files, then position the cursor on the sort to be opened.
3. Choose OK by moving the mouse cursor to OK and clicking the left button, or by pressing Enter. (You can also double-click the mouse button while the mouse cursor is located on the sort name.)

You can also choose the Open icon on the toolbar to access the Open dialog box. The Open icon is the first icon from the left on the toolbar. The sort is brought to the screen where you can make revisions. Whenever changes are made to a sort, save the sort again to save the changes. If you do not, you will lose all your changes.

Exiting *FastPro* and Lotus™ or Quattro™

When you are finished working with ***FastPro*** and have saved all necessary data and information, exit ***FastPro*** by choosing File, then Exit or pressing Alt + F4. You can also exit ***FastPro*** by positioning the mouse cursor on the Control-menu box (the upper box with a dash in it in the upper left corner of the screen), clicking the button once, and choosing Close in the menu that displays. After exiting ***FastPro*** for Lotus/Quattro, you must exit the Lotus/Quattro program. When you exit ***FastPro***, the Lotus/Quattro program group window appears on top of the Program Manager window. You should always close ***FastPro*** before exiting from the Lotus or Quattro platform. To do so, click on the Control-menu box in the upper left corner of the program group window, then choose Close in the menu that displays on the screen, or press Crtl + F4. If ***FastPro*** is running on your Lotus or Quattro platform in Windows, you exit Windows by completing the following steps:

1. Choose **F**ile, then **E**xit Windows or press Alt + F4.
2. At the Exit Windows dialog box, choose OK (or press Enter).

If you are using Windows 3.0 or higher, the Exit Windows dialog box contains the **S**ave **C**hanges option. By default, this option should have an X in the check box. With this option selected, Windows is exited and any changes made to the layout of Program Manger are saved.

Deleting Cells, Rows, and Columns

Remember that, when working in ***FastPro,*** if you select the contents of a cell, row, or column, and press Delete or Backspace, only the text is deleted. In order to remove the entire cell, row, or column, you must use the commands in either the Table menu or the Edit menu.

To delete a row of cells, complete the following steps:

1. Select the row to be deleted.
2. Choose **E**dit, then **D**elete Rows.

Or,

1. Locate the cursor anywhere within the row to be deleted.
2. Choose **E**dit, then **D**elete Cells.
3. Choose Delete Entire **R**ow in the Delete Cells dialog box.
4. Choose OK (or press Enter).

To delete a column within a sort, complete the following steps:

1. Select the column to be deleted.
2. Choose Edit or Table, then Delete Columns.

Or,

1. Locate the cursor anywhere within the column to be deleted.
2. Choose **E**dit, then **D**elete Cells.
3. Choose Delete Entire **C**olumn in the Delete Cells dialog box.
4. Choose OK (or press Enter).

To delete a cell or group of cells within a sort, complete the following steps:

1. Select the cells to be deleted.
2. Choose Edit, then Delete Cells.
3. In the Insert Cells dialog box, specify how you want the remaining cells shifted (up or left), then choose OK (or press Enter).

You could also select an entire row or column and then use the Cut command in the Edit menu. However, if you have selected a group of cells within a row or column and you access the Cut command, only the text within the cells will be deleted.

Merging and Splitting Cells

There will be numerous times, while working within sorts, that you will want to combine data by merging a group of cells in a row together to create one large cell. This is especially useful in creating a subheading that spans the width of a sort, for example. Using the Merge Cells command in the Edit menu, you can merge cells within the same row together.

When you merge a group of cells together within the sort, each original cell becomes a new line in the merged cell. Therefore, if you merge four cells together you will have one cell that is four lines in height. You should merge cells before entering text in them. Otherwise, when the cells are merged, the text within the original cells will be merged into a series of lines in the merged cell.

In ***FastPro***, only cells within the same row can be merged. You cannot merge cells from different rows together. For example, to merge all the cells in the top row of a sort to contain a title or subheading, complete the following steps:

1. Select the entire first row of the target row you wish to expand.
2. Choose Edit, then Merge Cells.

You do not have to merge all the cells in a row together; sometimes you may want to merge only some of the cells within a row. To do this, select those cells that you want to merge, then access the Merge Cells command in the Edit menu. If you make a mistake in merging cells together and wish to undo the merge, access the Undo Merge Cells command in the Edit menu immediately after merging the cells. You can also split cells that have been merged back into their original configuration by accessing the Split Cells command in the Edit menu. To do this, complete the following steps:

1. Select the cell to be split.
2. Choose Edit, then Split Cells.

The Split Cells command is not always visible in the Edit menu; the Merge Cells and Split Cells commands are toggle commands (they switch back and forth in the menu). If the Split Cells command is not visible, it means that you have not selected a cell that was previously merged (and you cannot split cells that were not previously merged).

Changing Column Widths

When a sort is created within ***FastPro***, the columns are the same width by program default. The width of the columns depends on the number of columns as well as the spreadsheet margins. In some sorts, you may want to change the width of certain columns to accommodate more or less text. There are three ways to change the column width in ***FastPro***:

1. Use the mouse to drag the column borders to the desired location.
2. Use the mouse to drag the column markers to the desired location.
3. Use the Column **W**idth command in the **E**dit menu.

To change the column width with the column borders, complete the following steps:

1. Position the mouse cursor on the column border to be moved. When the mouse cursor is positioned directly on a column border, it changes to a double-pointed arrow with vertical lines in between.
2. Hold down the left mouse button, then drag the border to the right or left to widen or shorten the column. (As you drag the border, the column border in the original location displays in blue, and the column border being moved displays as a faint gray line.)
3. When the column border is in the new location, release the mouse button.

When you move a column border in this manner, any column borders to the right move the same distance in the same direction. In this way, the columns to the right of the border you are shifting remain their original widths. If you want to move a column border without moving the columns to the right, hold down the Shift key while dragging the column border to its new location. In this way, the location of the columns to the right is not changed, but the width of the column immediately to the right is resized. Any other columns to the right of the border you have shifted remain their original widths. If you want to move a column border without moving the columns to the right, but want those columns to be resized proportionately, hold down the Ctrl key while dragging the column border to its new location.

You can also use the Column Width command to determine the measurement of a particular column in a datatable. To do so, position the cursor in any cell within the appropriate column and access the Column Width dialog box. The current measurement for the column width appears in the appropriate text box. With the Column Width dialog box displayed on the screen, you can also enter specifications for the previous column and the next column by clicking on these buttons first, then entering the appropriate specifications.

Changing Row Height

The height of a cell is determined by the amount of text, and its point size, within it. The height adjusts automatically as the text wraps around from one line to the next within the same cell. The height of an entire row is determined by the tallest cell in the row; all cells adjust to be the same as the tallest cell.

In some cases, you may want to establish a minimum row height, independent of the amount of text contained in the cells. You can create a variety of special effects by varying the row height. For example, you might want to add more blank space between the title of a datatable and the surrounding border, or separate one row of cells from another by changing the row height of one of the rows. To change the row height, you must locate the cursor in the row you wish to change, then access the Row Height command in the Edit menu. When you access this command, the Row Height dialog box appears allowing you to set the row height to your desired setting.

Changing Cell Alignment

When you initially create a sort, the alignment is pre-established in ***FastPro*** by default. If the cursor was located in a text line, paragraph, or data table that was centered when you inserted the data into the sort, each cell in the text line, paragraph, or data table will use center alignment. In this case, the end-of-cell markers appear in the center of each cell.

To change the alignment of cells, select the cells, then choose the desired alignment button on the toolbar, or access the paragraph command in the Format menu. Different alignment options can be applied to different cells.

Printing

When a sort is created in the computer, gridlines display on the screen identifying individual cells within the sort. These gridlines will not print when the sort is printed, unless you choose to do so by selecting Grid Lines in Print Preview. This is the default setting in ***FastPro;*** however, the network timeline should always be printed with the grid lines to provide easy visual alignment of the multiple dates.

In printing the sorts, remember ***FastPro CPM Scheduler*** software is programmed in PostScript fonts and will reproduce best with a laser printer. Acceptable sort copies can be printed on a dot matrix printer, but some of the finer graphic artwork on the sorts will not appear with dot matrix printers.

Using *FastPro* in Windows™

FastPro operates on your Lotus™ or Quattro™ spreadsheet program within the Windows environment created by the Windows program. However, when working in ***FastPro***, the term *window* refers to the screen, excluding the menu bar, status bar, and message bar. A window therefore contains the portion of the screen where you enter or edit text, formulas, the title bar, and the scroll bars.

The Windows program creates an environment in which various software programs are used with menu bars, scroll bars, and *icons* to represent programs and files. With the Windows program, you can work on several different software programs at one time. Similarly, using windows in ***FastPro***, you can work on several different sorts at one time.

With multiple windows open you can work on several sorts, or two different parts of the same sort at once. You can copy or move information from one window to another, or compare the contents of several different windows. ***FastPro***'s windows feature is very useful for:

- Looking up information in one part of a sort while working on another part of the sort.
- Copying or moving text or formulas between different sorts or different parts of the same sort.
- Comparing the contents of several different sorts at once.

Opening A Window

The maximum number of windows you can have open at one time is nine. Therefore, you can work on up to nine different sorts at one time. When you open a new window it is placed on top of the original or preceding window. Once multiple windows are opened, however, you can resize the windows to see all or a portion of them at one time on the screen. There are three ways to open a new window:

1. Use the New Window command from the Window menu. This command opens a new window, which is placed upon the original window.
2. Use the New command from the File menu. This opens a totally new window.
3. Open another file (sort) with the Open command from the File menu. Use this command to open a different file (sort). The new file will be contained in a new window, placed on top of the original window.

At the bottom of the Window menu is a list of all open windows. The name with the checkmark next to it indicates the activity window being used (the one in which the cursor is located).

Moving between Windows

Once you have opened a second window, the original window is hidden behind the new window. To move back and forth between windows, you can choose one of the following methods:

1. Press Ctrl + F6 for the next window, or Shift + Ctrl + F6 for the previous window.
2. Choose Next Widow in the Document Control menu.
3. Access the Window menu and select the appropriate file name on the menu.
4. Click on the title bar of the appropriate window (when it is visible).

When you move back and forth between windows in this manner, the cursor in each window remains in the same location as when you last worked with it.

Copying and Moving between Windows

The most common use of multiple windows is moving or copying text and formulas between windows. To do this, use the range operation explained in

Data Range Commands, together with the information presented in this chapter. For example, to move text or formulas from one window containing one sort to another window containing a different sort, complete the following steps:

1. Select the text or formulas you wish to copy or move.
2. Choose Edit, then Cut (or Copy), or use one of the other methods for transferring data to the clipboard.
3. Open the sort (file) in which you wish to insert the text (or locate the cursor in the appropriate window).
4. Position the cursor in the desired location.
5. Choose Edit, then Paste, or use one of the other methods for pasting information from the clipboard.

The steps are almost the same as when moving and copying information within the same sort. The advantage of working with multiple windows, however, is that you can transfer information *between* sorts very easily.

If you are using Lotus for Windows, version 3.0 or higher, the steps involved in transferring information from one window to another can also be used to transfer information from one pane to another in a split window.

Moving Windows

Once you have multiple windows open you can move them around on the screen without changing their size. This is particularly helpful when working with several different small windows at once. To move a window, you can use the Move command in the Document Control menu, or use the mouse to drag the window to the desired location. To move a window, complete the following steps:

1. Choose the Document Control menu, then Move. If you are using the keyboard, press Alt, -, M; or use the speed command Crtl + F7. (If the Move command is unavailable, choose Window, then Arrange All first.)
2. Press the arrow keys continuously to position the window in the desired location.
3. Press Enter to confirm the move.

If you are using a mouse, you can move a window easily by positioning the mouse cursor on the title bar of the window you want to move, then holding down the left mouse button as you drag the window to the desired location.

Closing Windows

FastPro offers several different methods for closing windows. To close an open window individually, you can do any of the following:

- Choose File, then Close.
- Choose the Document Control menu, then Close.
- Press Ctrl + F4.
- Double-click the Document Control box at the left side of the title bar.

If you access the Close command in the File menu when you have more than one window open that contains the same sort (if you selected the New Window command in the Window menu), all windows containing that sort are closed at once. If you have not saved your changes before closing a window, ***FastPro*** prompts you to do so.

You can also close all open windows and exit the ***FastPro*** program at the same time. To do this, access the Close command in the Applications Control menu (represented by the gray box at the left side of the application title bar) by pressing **A**lt, space bar, Close, or by pressing Alt + F4.

Maintaining Schedule Sorts

Every project schedule maintains a filing system. The system may consist of documents, folders, and filing cabinets, or it may be a computerized filing system where information is stored on disks. Whatever kind of filing system a project uses, the daily maintenance of files is important to the project schedule's successful completion. Learning how to maintain sorts saves a great deal of time in searching for particular information and avoids costly mistakes from accidentally deleting or writing over existing files.

Since ***FastPro*** operates from a computer's hard disk, it saves information automatically to that hard disk unless you specify another drive. Effective file maintenance is particularly important on a hard disk; however, it also applies to floppy disks. Maintaining files on a disk usually includes such activities as deleting unnecessary documents and files, copying important documents and files so that you have a backup available, and establishing a logical way of organizing where files are stored.

A hard disk is usually divided into several different subdirectories. You can think of subdirectories as file folders. File folders contain documents that have something in common; so do subdirectories. There are many different ways of organizing computer files into subdirectories. Some of these include:

- ❑ By the program in which they were created
- ❑ By the name of the person who generated the files
- ❑ By the name of the person who input the files
- ❑ By the type of file (sorts, memos, letters, etc.)

Subdirectories are most commonly found on hard disks. However, a floppy disk can be divided into various subdirectories also. To choose a different subdirrectory in the Open dialog box, access the Directories option, then select the appropriate directory from the list box. To choose a different drive, access the Drives option, then select the appropriate drive from the drop-down list that displays. To open a sort that is stored in a subdirectory on the C: drive, complete the following steps:

1. Choose **F**ile, then **O**pen.
2. Choose **D**irectories, then select the root directory (C:\) in the Directories list box, and choose OK (or press Enter). (If you are using the mouse, you can double-click on the C:\ at the top of the Directories list box.)
3. Select the desired subdirectory in the Directories list box, and choose OK (or press Enter). If you are using the mouse, you can double-click on the desired subdirectory instead.
4. Choose File Name [un], then select the sort or file you wish to access in the file list box, then choose OK (or press Enter). If you are using the mouse, you can double-click on the sort or file name instead.

Opening the same sort or file from a floppy disk is a somewhat simpler procedure, since you would not normally need to specify a subdirectory on a floppy disk. To open the desired sort or file from the disk in drive A, complete the following steps:

1. Choose **F**ile, then **O**pen.
2. Choose **Drives**, then select a: in the drop-down list that displays, and choose OK (or press Enter). If you are using the mouse, you can double-click on a: instead.
3. Choose **File Name**, then select the desired sort or file from the file list box, and choose OK (or press Enter). If you are using the mouse, you double-click on the sort or file name instead.

Saving Sorts To A Different Subdirectory

Unless you specify otherwise, ***FastPro*** automatically saves every sort you create in the Windows\Lotus (Quattro) subdirectory on the hard disk. You may, however, want to save a sort to another subdirectory. To do this, you must access the Save As dialog box, then change the information in the Directories line at the top of the dialog box.

In order to change the information in the Directories line, follow the same procedure as when you opened a sort from another location. For example, if you want to save a sort to the Sort By I-J Numbers subdirectory on the C: drive, complete the following steps:

1. Choose **File**, then **Save As**.
2. Choose **Directories**, then select the root directory (C:\) in the Directories list box, and choose OK (or press Enter). (If you are using the mouse, you can double-click on the C:\ at the top of the Directories list box.)
3. Select the Sort By I-J Numbers subdirectory in the directories list box, and choose OK or press Enter. If you are using the mouse, you can double-click on the Sort By I-J Numbers subdirectory instead.
4. Choose **File Name**, then key in the sort name.
5. Choose OK (or press Enter) in the Save As dialog box.

Saving Sorts To A Different Disk Drive

Any sort can be saved to a floppy disk by entering certain instructions in the Save As dialog box. There are many reasons for saving documents to floppy disks. You may want to make a backup copy on a floppy disk, you may need to take the sort elsewhere to be printed into hard copies, or you may intend to mail the disk to another location. In any case, you can save a sort to a different disk drive in much the same way as you open files from a different disk drive. In order to make a backup copy of the file named Sort By I-J Numbers and save it on the disk in Drive A, complete the following steps:

1. Choose **File**, then **Save As**.
2. In the Save As dialog box, choose **Drives**, then select a: in the drop-down list, and choose OK (or press Enter). If you are using the mouse, you can double-click on the a: in the drop-down list instead.
3. Choose OK (or press Enter) in the Save As dialog box.

Unless you plan to change the name of a sort when you save it in another location, you do not need to key in the sort name again. ***FastPro*** uses

the same sort name, but stores the sort in the location that you specify. Your sort can be saved to any drive in addition to, or instead of, the C: hard drive.

Chapter

16

Project Scheduler Consultant

Running the Shop

There are echelons within echelons, circles within circles. The big picture can usually be analyzed by looking at smaller similar pictures. Many of today's successful project scheduler consultants share common traits. Typically, those traits tend to be:

- An eye for opportunity. Many successful project schedulers can sense the needs in the cyclic nature of the local construction industry, begin to see where past trends predict that future needs for projects will develop, and then move quickly to service clients in that area. This could be either an entirely new business, such as the semiconductor industry"s growth, or simply filling the need of present clients' expansion plans.

- An appetite for hard work. Most professional project schedulers who own their own consulting business work long, hard hours. They tend to focus on real priorities, especially when times are tough. Those who avoid the tough problems by burying themselves in their original area of expertise and not keeping up with industry change often fail.

- Discipline. You can work hard, or you can work smart. It's not how hard you work, but the way you work that counts. Successful consultants resist the temptation to do what comes easiest. Instead they do what is most essential to the growth of their whole business.

- Independence. Successful project scheduling consultants have a strong drive toward independence. Conversely, they also have refined abilities to be a good project team player when the need for cooperation arises. They can become managers of other people's work so their business can grow quickly and achieve substantial success.

- Self-confidence. Consultants must be extremely self-confident to accept the risks involved in starting a business. This helps them overcome enormous obstacles to achieve their goals. In addition, they can temper their self-confidence with objectivity and change course to cut losses where necessary.

- Adaptability. Successful consultants adjust quickly to the ever-changing demands of a growing business within a demanding industry, and develop the skills to deal with those changes. Change is one of the driving motivators in all modern businesses and the construction industry is no exception.

- Judgment. Consultants make decisions based on experience and intuition, without taking too much time to make up their minds. When they lack the knowledge to make sound decisions on their own, they seek information and advice from other professionals without hesitation.

- An ability to tolerate stress. Schedulers face many crises on the road to prosperity. To survive, they often unwind with sports and hobbies. Many are encouraged by the understanding and support of their spouses and families.

- A need to achieve. Successful project scheduling consultants have a single-minded, almost compulsive desire to succeed. Although financial rewards are important, they are often secondary to the person's drive toward personal achievement.

- A focus on profits. Running a consulting business brings many rewards like independence, community prestige, and personal satisfaction. But successful project scheduling consultants never lose sight of the need to make a profit. They know that return on investment for their clients' projects is the ultimate benchmark of success.

- Self-awareness. Almost all consultants start their business with some weaknesses. Those who succeed recognize their limitations and are always ready to seek and use outside help when the situation demands it.

As an outside services consultant, your principle business concern is to focus on current construction technology, educating yourself to learn the available quality compromises, developing workarounds to circumnavigate bottlenecks in production, and researching methods of improving quality project scheduling in the future. Demanding clients, increasing defect litigation, and a very competitive market mandate professional-quality CPM scheduling. A good CPM project schedule will define the plan of production in a clear network diagram, provide guidance for the activities subcontractors, and present the goals and specific objectives of the construction phases in time scaled milestones. Focused attention on every detail makes the difference between success and failure for the owner of a fledgling consultant business. While the fast-tracking design/build project delivery method will be with us for the next millennium, there are many other futuristic types of scheduling systems already on the horizon. Compiled management, just-in-time project execution, and "bridging" types of project scheduling are currently in their infancy stages, but if they prove to be viable and cost-effective systems you can expect them to take off quickly and leap to the forefront, as did fast-tracking in this decade. New technologies are also upon us and in use now. The only things holding back the use of graphite composite plastics in structural construction components are the code restriction imposed by fire marshals, who are understandably concerned about plastic offgassing in a fire. When some enterprising scientist invents a way to inhibit composite plastics offgassing in fires, we'll be using the stuff by the truckload the next day.

Additionally, in the last two years steel framing has made great strides in the residential construction industry. As the lumber industry continues to become more volatile in the future, using steel in residential construction instead of wood will continue to grow. Other things to look for and take advantage of as the future speeds ever closer are standardization of engineering and building codes; new developments in fastener and connectors systems, tools, and materials; and new products and construction techniques.

The key elements in building a project scheduling consultant practice can be boiled down to these:

- Credential yourself. Become the best you can be. Know your local field and do a good job for your clients.
- Build contacts through existing clients. Build strong personal relationships with whomever you deal with in a company.
- Start networking with other people in the company. Join business groups. Get to know the individuals who might recommend you.
- Follow up. You can't just have lunch and walk away, thinking you'll get the business. You have to ask for it and continue to keep in touch, sending out information and keeping your name in front of the potential client.
- Once you get the business, nurture it. Return all phone calls. Keep in touch.

Tax Benefits

When you own your own business, you're entitled to many special financial perks with full IRS approval. Once you know the rules and your business starts operating by them, the IRS will allow you to use before-tax dollars to pay for many things. You may even reduce your personal expenses, by reclassifying them as legitimate business expenses. Here are examples:

- Deduct for home office (see subsection following).

- Drive a luxury car. As a business owner you get tax breaks on the business driving you do. This covers fuel, oil, repairs, maintenance, insurance, licenses, washing, and several thousand dollars in depreciation, whether it's for a compact or luxury car. You choose which one it's to be.

- Visit resort locations. Attend a business convention or trade show in a desirable resort location in the U.S. All your business-related travel and living expenses (including staying at posh hotels and fancy meals) are deductible, even though you play tourist in your off hours.

- Take the family along. If your spouse and family are involved in the business, your business can pay for them too.

- Free vacation travel. When planning any business trip, consider extending it with some vacation time. Providing the primary purpose of the trip is business, you can deduct the round trip cost to your business destination, even though you take time off to vacation before or after the business event.

- Feel like a million dollars. You can rent any car on a business trip and deduct 100% of the expenses, whether it's a compact or an exotic sports car. One tip is to rent the model you're considering buying, if that's appropriate. Then you can give it an extended test drive as a bonus.

- Move up to first class. As the boss, you decide how you prefer to travel. Sure, the difference between super saver and first or business class fares is considerable, but since both are fully deductible it costs you less to travel in style.

- Take a $2000 cruise. Attend a convention on a cruise ship registered in the U.S. and with ports of call in the U.S. or its possessions to actively conduct business, and you will enjoy all the pleasures of life on board, with up to a $2,000 deduction per year.

- Continuing education in your own field is usually fully deductible. This includes reference manuals and software such as this book, seminars, tuitions, trade associations, periodicals, degrees, trade fairs, etc.

As a consultant, your business operations will consist of scheduling multiple projects simultaneously, sharing the data and sorts with members of the project teams in a timely manner, and creating convincing reports, spreadsheets, and graphics for clients. You'll do this by organizing your communication with the project teams, controlling the schedules, and cost-tracking by audit trails.

Home Office Deduction

As you build your business, you may be thinking about the tax deductibility of the space you have set aside for work done at your home. If you run your business out of an office that is not located in your home, you will have difficulty claiming you have two offices, one in your home and one at another location.

But if you are running your business from a home office only, consider the following: First, remember that much cheating goes on in this area. Many people use the laws to claim deductions that are not legitimate. Thus, the IRS is cracking down on misuse of the home-office deduction. However, if you are legitimately running your business from your home, you are entitled to certain deductions.

You can write off some home expenses if you use a portion of your house regularly and exclusively for business and if it is your principle place of business. Your office can be part of another room, but it must be maintained as a permanent work area. Your living room coffee table or dining room table cannot be your "office" for tax purposes if it is set up as an office only a portion of each day.

Calculating Your Office Deduction

To determine the amount of your basic deduction, figure the percentage of your house that you use as an office, in square feet. Let's say that it is seven percent of the total square footage of the home. Then you can deduct up to seven percent of your mortgage interest, rent, taxes, homeowner's insurance, and building upkeep such as exterior painting. You can also take a depreciation deduction for the space that is set aside for your office.

Of course you cannot write off more of your home expenses than you produce in home-based business income. If you make no sales this year, you cannot take a deduction.

Other Office Expense Deductions

Other deductions can be taken for supplies, equipment, business insurance, your accountant and CPA fees, etc. Be sure to also keep your records and receipts for those home expenses that overlap your home-based business for a minimum of five years, and all tax records for a minimum of seven years. Examples of these expenses are: Your business uses postage stamps which are a business expense, but you and your family also use postage stamps and should not use those stamps purchased by and for the business for personal use. Keep the receipts for the purchase of stamps for the business, as well as receipts for personal stamps. At the time of a tax audit, receipts for personal items will be valuable, showing you have not been trying to deduct personal items as if they were business expenses.

New to the tax code for 1996, you can deduct up to $17,500 in business-related equipment purchases, but not more than the business earned. Before 1995, the limit for this deduction was $10,000.

Phone Deduction: The IRS will not allow a business deduction on the first telephone line coming into your home, even if it goes to the telephone in your home office. However, long distance business- related calls are deductible on that first line. A second line coming into the house is deductible, if used exclusively for business. Additionally, if you are careful to make only business calls on your business line and only personal calls on your personal line, the business line's billing statement makes bookkeeping and tax reporting easy.

On a related subject, most self-employed people fail to realize the importance of sheltering income. Every business person should open an Individual Retirement Account (IRA) and contribute as much as possible to it during the year, up to the $2,000 current limit. See your accountant or banker regarding opening an IRA account.

Building A Clientbase

In any business, one must have an adequate customer base. Where does the project scheduling consultant look to build that customer base? Working for others until one has built a reputation, then start an independent business is the traditional success story. Personal contacts represent the single largest source of business for new consultants. Small to medium-sized consulting businesses rely on referrals for income. The use of advertising in the yellow pages, and direct mailings to developers and companies who contract for their work, is recommended in business management courses and by the SBA. However, it is always important to remember that seasoned businesspeople see advertising as a black hole you throw your money down and hope to get some return on. For established consulting businesses, advertising is barely cost-effective at best, and for new companies it's a total loss. (Great... Now I've ticked off the advertising people. My editor is going to beat me with a stick when she reads this. Maybe if you write in, they'll send you some free pictures of the event.)

But seriously, think about it for a second. How many business consultants are running 60-second commercials on your local cable station? Exactly the same number who ran display ads in last sunday's paper. Zero. Personal networking is what builds a customer base for new consultants. Save your advertising dollars for shoe leather and get out there and pound the streets. Knock on some doors and make some contacts. After you've made your first million, you can hire an ad agency to put your name up in lights.

Once your business does start to grow you can develop a formal marketing plan to market your business. Lack of a marketing plan will kill a growing business, regardless of how strong the other areas of the business may be. On the other hand, you need a solid overall plan, too, because too many sales will take an unprepared business under just as quickly as too few.

Generating clients can start the moment you employ fundamental business management procedures in client analysis and in identifying the demand market. Everything in every business begins with the customer. Therefore you begin by identifying the consumer of your service. Who are your clients? You need to know as much about them as possible in terms of typical projects they handle, what their financial track record is, and what their potential for future projects will be. In terms of what their business may bring to your business, you need to get inside their heads about what they build and how they make their profits from those projects. You will find as your career advances that the ways of making money through real estate development are quite varied.

At a low level of specificity, you can hypothesize (guess) about these issues. At a higher level of specificity, you study them through secondary research, which is information and data already gathered about typical project owners like your potential clients, then primary research which consists of surveys and studies directly related to those specific potential clients. The "movers and shakers" in the construction industry within your local area are probably already known to you, or will be soon enough after some detective work on your part. Once you've done your homework, you use your data in two principal ways:

- Communicating your presence and offering your services to them efficiently. What business network do they use to get funding? What attracts their attention? Are they looking for a new scheduling company or competitive bids on service? Have they lost money with inefficient or non-CPM-trained project schedulers? Are they looking for someone like you for their next project? You must make your presence known to them in a crisp and professional presentation.

- Developing your consulting service to best meet their needs. What aspects of your scheduling skills will most appeal to them? CPM? Fast-tracking? Cost control? Do a little role reversal here. Once you've identified a developer, if you were that developer, what would you be looking for in a local project scheduler? Become that professional.

The next step is segmentation. To segment your clients means to divide your client base into useful groups. You would like to find natural divisions like age, types of projects preferred, funding capabilities, areas of interest, etc., which allow you to focus your target market efforts. Then developing presentation strategies for targeted clients will make it easy (and economical) for you to access them as a group. Start with some suppositions about the groups of clients. Then fill in the details. As you proceed, think about what you would most like to know and about how you might collect information about your potential clients in the future. Check your research and estimate data for accuracy by comparing information to statistical sources such as builders' exchanges, trade associations, labor unions, trade journals, national SIC statistical compilations, government research, area construction research, and general or specific readership publications.

Finally, compare your data to a "primary market study." The study and its analysis will be a substantial undertaking. Be sure you are asking the right questions. You may be able to do the actual client surveying yourself, but unless you are experienced in such surveys you should use a market study professional to develop the survey documents of the recent construction projects in your area and process the survey for your target market. For any standard consultant business, these primary market studies for target markets include all of the following segmentations:

- Distinguishing the characteristics of your primary target market client and market subdivisions. Narrow the number of targeted clients to a manageable size. Efforts to penetrate target markets that are too broad are often ineffective. Begin by identifying:

    a. Critical client needs
    b. Extent to which those needs are currently being met
    c. Demographics
    d. Past projects locations
    e. Project team decision makers and influencers
    f. Seasonal and cyclical trends

- Next determine your primary target market size
    a. Number of prospective clients
    b. Annual projects requiring the same needs as your service
    c. Local areas of construction growth
    d. Anticipated construction market growth, short and long-term

- Measure market penetration and determine the extent to which you anticipate penetrating your competition's existing market. Include reasons why your company's penetration is achievable, based on current market research.
    a. Market share
    b. Number of current clients held by total local market
    c. Area coverage of target market base

- Fee pricing and gross margin targets
    a. Competition price levels
    b. Competition gross margin levels
    c. Client discount structure (volume of business, prompt payment, etc.)

- Methods by which specific clients within your target market can be identified
    a. Developers directories
    b. Trade associations
    c. Builders exchanges
    d. Research data

- Methodology through which you will contact specific members of that target market
    a. Personal networking
    b. Publications
    c. Direct mail brochures
    d. Sources of influence and advice

- Hiring cycles of your prospective clients
    a. Needs identification
    b. Research for solutions to needs
    c. Solutions evaluation process
    d. Final solution selection responsibility and authority (owners, developers, project managers, engineers, etc.)

- Key trends and anticipated changes within your local construction industry target market

- Secondary target markets and key attributes

a. Needs
b. Demographics
c. Significant future trends

Follow-up results need to be divided into manageable categories, such as:

- Potential clients contacted and follow-ups
- Information and presentations given to potential clients
- Reaction of potential clients
- Importance of satisfaction of targeted needs
- Test groups of clients' willingness to use your services at various fee levels

Then a comprehensive survey of your competition must be done that includes:

- Identification
  a. Existing
  b. Their market share
  c. Potential (how long will your "window of opportunity" be open before your initial success breeds new competition? Who will your new competitors be?)
  d. Direct
  e. Indirect

- Strengths (competitive advantages)
  a. Ability to satisfy clients' project scheduling needs
  b. Market penetration
  c. Track record and reputation
  d. Staying power (financial resources)
  e. Key personnel in your local business network
  f. Personal resources

- Importance of your clients' business to your competition

- Barriers to entry into the market
  a. Cost (investment)
  b. Time
  c. Technology
  d. Key personnel
  e. Client inertia (company loyalty, existing relationships, etc.)

Identify that target market by:
a. Personal networking
b. Publications
c. Direct mail brochures

d. Sources of influence and advice

Then start a hiring cycle of prospective clients through:

a. Needs identification
b. Research for solutions to needs
c. Solutions evaluation process
d. Final solution selection responsibility and authority (owners, developers, project managers, architects,engineers,etc.)
e. Taking advantage of key trends and anticipated changes within your local construction industry target market

Business Management

The legal organizing of your business begins the process. Establishing your business is the first step in business management. This is the legal form of your business that determines the available sources of financing, the extent of personal liability you have in the business, the extent of management control you have, and the company's tax liabilities. The legal form of your business may change as the business grows and should be review as financing requirements change. The sole owner often lacks the personal capital resources necessary to sustain growth. The partnership is usually created to expand the source of funds. Continued growth and the resulting larger financing needs may lead to incorporation which, in turn, provides access to external funds.

Sole Ownership is the first type of business organization. This is personal ownership, often called sole proprietorship, and relies primarily on the financial resources available to an individual. The owner has sole responsibility and complete control. He or she must obtain all the financing and is personally liable for any claims against the business. The sole ownership is easy to set up and is subject to minimum regulations. Business income is reported as personal income. Sole ownership appeals to individuals who value smallness, simplicity, and personal control.

General Partnership uses the financial and personal resources of two or more individuals, who share in the owning and running of the business. It is easily established, requiring only registration of the business name. A lawyer should be consulted to formalize the rights and responsibilities of the partners regarding management and profit sharing. The partnership agreement terminates with the death or withdrawal of a partner or the addition of a new partner. The partnership is not an entity separate from the partners. Each partner is personally liable to the extent of his or her personal assets. An individual's share of the partnership income is taxed as personal income.

Corporation: As a legal business organization, a corporation is a separate legal entity apart from the owners which has been created by the government. It can make contracts, be held legally liable, and is taxed as its own entity. The corporation may raise capital by selling stock to private investors or to the public. Stockholders are usually not liable for claims against the corporation beyond their own investments. Creditors may claim only corporate assets, although the corporate officers may be held personally liable. The corporation has no fixed life. The death of a stockholder or sale of one's personal investment will not disrupt the business. Corporations pay a graduated federal tax, as well as a state franchise tax to the Franchise Tax Board, for the privilege of doing business.

Taxes and Permits. The owner of a business is required to pay taxes and fees at the federal, state, and local levels. If the business is a sole ownership or partnership, estimated federal and state income taxes must be prepaid or paid quarterly. A Federal Employer's Tax Identification Number must be requested from the Internal Revenue Service, and federal income and social security taxes must be withheld from employee wages. Contractors must register with the Employment Development Department and pay unemployment insurance tax for employees who earn over $100 in a calendar quarter. California requires a seller's permit for businesses that sell tangible personal property. The contractor must collect the required sales tax from the customer unless the tax on materials was already paid at the time of purchase.

Purchasing materials wholesale and collecting sales tax requires a resale permit from the Board of Equalization. Required city and county project permits must be obtained from city or county clerk's offices before starting work.

Bonds: A bond is simply an insurance policy that has been purchased to guarantee that an event will either occur or not occur. A requirement of bonding is generally mandatory for large projects financed by institutional lenders. In addition, many developers, owners, lenders, and prime contractors impose bonding requirements. Bonds can be obtained from bonding companies for a percentage of the contract price, usually in the 1% to 2% range. This requirement is a cost of doing business and should be recognized when the bid is submitted. The project scheduler should affirm these bonds are in place before assigning critical activities workaround responsibilities.

Performance bonds: Guarantee the project's completion according to the building plans and specifications. if the job is abandoned or the work is unacceptable, the bonding company has the option of hiring another contractor to complete the work or of settling for damages.

Payment bonds: Guarantee that a contractor will pay his subs and suppliers, and assure the project's owner that no liens for labor and materials will be filed against the property.

Contract bonds: Guarantee both job completion and payment of all labor and materials.

In general, the bonding company will not have to pay more than the face amount of the issued bond. Bonding requirements may exclude a new contracting business from bidding on desired jobs. Bonding companies will not take the risk without verifying the technical and resource capabilities of the bonded contractor. If you are a contractor, it is essential then for your business to practice sound business management techniques if you hope to be able to qualify for bonding.

Accounting

Accounting Control

Accounting has frequently been called the "language of business" because accounting is a method of communicating business information. While financial statements often appear to be complicated, they contain the information necessary for planning, coordination, and control of your consulting business. The success of your small business will depend on your ability to react to change quickly enough to make the right move at the right time. Financial statements provide the basic tools for this analysis. The foremost objective of the accounting process is to arrive at an estimate of the periodic income and expenses of a business. With this information, the business owner can make the decisions which will increase the potential for profit.

Records

Every business must keep records. Running the business out of your personal checkbook is never sufficient. Good records are essential for effective management and required when seeking outside financing. In some cases, recordation is mandated by law. State and federal agencies frequently specify certain records which must be maintained and made available for government audits and to substantiate tax reporting. Accurate records must be kept to provide the foundation to construct financial statements. Information is summarized in these reports that highlight the total operational costs, individual job costs and other costs which affect your ability as the business owner to make a profit. Accurate records also provide information on cash receipts and a basis for the control of disbursements. The business owner can, by reviewing these records, determine cash flow requirements of other jobs which he or she has completed to help in estimating the cost of future jobs. In project scheduling, the most accurate data you can use for factoring comes from your company's historical records.

Cash Versus Accrual

Under the cash basis of accounting, revenue is recognized and the transaction is recorded when the money is received, and expenses are recorded when they are paid in cash. Since final profit cannot be determined until after the work is completed, the cash basis of accounting is not recommended for contractors or consultants. The accrual basis of accounting is a method that recognizes revenues as earned and the transaction is recorded when the work is complete. Expenses are recorded when incurred, regardless of when payment is received or made. Money due the company is recognized as an account receivable (AR) and the value is shown as an asset before payment is received. This method is also a part of the double-entry bookkeeping system because every journal entry requires a counterbalancing reciprocal entry in the opposing column to balance out the financial statement.

The Bookkeeping Process

Your Financial Statements

Once you have chosen your method of accounting and have obtained the services of a bookkeeper or accountant, daily transactions are recorded in a journal on a daily basis. Periodically, the bookkeeper will transfer these daily journal entries into their appropriate summary reports, which are called ledgers. The bookkeeper will set up a "chart of accounts" containing all the accounts necessary in the particular business. Each account will maintain a balance much like a checking account balance. When a balance sheet is prepared, these balances will be transferred to it. Information from monthly balance sheets then is transferred to summary reports called income statements. The balance sheet and income statement summarize the firm's internal data. These statements, in turn, provide the information for ratio analysis, which highlights the strengths and exposes the weaknesses of the company. Some of these financial statements (boldfaced for quick reference), are further explained below.

Balance Sheet

The balance sheet is a statement of the financial condition of an individual business at any given moment. The income statement lets the reader compare where the company stands at the moment to how it stood last year and in previous years. The balance sheet itself is divided into two sides; on the left are shown the assets and on the right are shown the liabilities and owner's equity (stockholder's equity). In the double-entry system, both sides are always equal or in balance; hence, the name. The listed assets of the company include its physical goods and financial claims on others. Liabilities represent the claims of others, such as suppliers, on the company. These are called accounts payable (AP). The stockholders equity section includes the original investment of the owners, plus the undistributed profits earned by the corporation to date. These are the main balance sheet components:

Current Assets

Assets are categorized as either current or fixed. Current assets are listed first on the balance sheet. These are the assets which will be converted into cash in the normal course of business, generally within one year from the balance sheet date. In addition to cash (money on hand and deposits in the bank), more of the other items which will be turned into cash soon include:

Retention. This is an amount owed by customers, which is due and payable upon project completion and expiration of the lien period.

Accounts receivable. The amounts due from customers (other than retention) in payment for business services.

Inventory. Includes all materials, labor and overhead on jobs which are currently in progress, less any amounts which have been billed customers. Inventory also includes raw materials on hand which have not been assigned to any specific project.

Prepaid Expenses. Goods or services the company buys and pays for before use. Examples are insurance premiums and office supplies.

Fixed Assets. The next major category on the balance sheet is property, plant and equipment, sometimes called fixed assets or plant and equipment. These assets include physical resources a business owner has or acquires for use in operations, of which he has no intention of reselling. Fixed assets are valued at original cost less accumulated depreciation, regardless of the current market value. Sometimes property, plant, and equipment may be leased rather than owned. The value of the leased property is often included with the fixed assets, and the lease payment is included with the liabilities to balance out.

Other Assets. Resources not included under current assets or under property, plant and equipment (fixed assets) are placed here. Examples include excess materials or equipment held for resale and long-term receivables.

Liabilities. Liabilities, like assets, are broken down into two major categories: current liabilities and long-term liabilities. Liabilities represent obligations to pay money or other assets or to render future services to others. The relationships between current and fixed assets and current and long-term liabilities will become apparent as you learn how to analyze the information presented in your financial statements.

Current Liabilities. This item includes debts of the company which become due within one year of the balance sheet date. Current assets provide the source from which these payments are usually made. You must be aware of this relationship to maintain sufficient current assets or to control the amount of current liabilities to avoid becoming delinquent in your bills.

Accrued expenses payable. Monies owed for interest, services, insurance premiums, and other fees are not included under accounts payable. Thus, expenses that have been incurred but are not due for payment on the date of the balance sheet are grouped under accrued expenses payable.

Accrued payroll. Salaries and wages the business owner currently owes to employees.

Long-Term liabilities. Long-term liabilities are notes or mortgages due one year or more beyond the balance sheet date.

Stockholders' equity. The stockholders' equity section of the balance sheet, also called net worth, equity, or net assets, represents the claim of the owners on the assets of the business. Different business forms use different names for this section of the balance sheet, but basically this is the original investment of the owners.

Capital stock. The total amount invested in the business by the business owner in exchange for shares of common stock at par value. Par value is arbitrarily established and need not be the same as the current market price of that share of stock.

Retained earnings. The total earnings of the corporation from its start, minus the total dividends declared (distributions to owners) since the corporation was founded. This account represents additional investment by the owners, who were willing to forego a larger distribution of the company's earnings.

Income statement: The income and expense statement summarizes the operations of the company over a period of time, usually a fiscal year. In the construction industry, income statements are often prepared to cover shorter periods, e.g., a quarter or a month. The income statement is also called an income and expense statement, profit and loss statement, or abbreviated as a P & L statement. Basically, the income statement shows business revenue, expenses, and resulting profit or loss for a given accounting period, normally a quarter. Since the revenues are matched against the related costs and expenses, the difference between the two is how much the company makes or looses. This profit or loss is often called the "bottom line" because it is an important indicator of performance. Also, it is the last line of the income statement. However, it is only one performance indicator that financial statements report. The relationships of the various expense categories to total revenue presented in the financial statement must be understood if the company is to be effectively managed.

The components of a typical income statement are as follows:

Sales income. The figure for sales of services represents the revenue flowing into the consultant's or contractor's business from the sale of his service. This also reflects all the billings made to customers for completed projects, as well as work in progress.

Cost of operations. The cost of operations comprises all costs and expenses associated with running the business, with the exception that federal income taxes are shown on another line. By totalling all the costs of operations and the cost of goods sold, then subtracting these from revenue, the operating profit of the company can be established. This represents the amount of profit earned by the business owner without taking into consideration federal income taxes.

Cost of goods sold. Direct costs are those expenses directly chargeable to a specific project. This figure includes direct labor, direct labor burden, materials used specifically for that project, and other direct costs.

Direct labor. This represents the actual cost for all jobs worked on during the period covered by the income statement. The amount of direct labor and the percentage of direct labor to the total cost of any particular project should be closely monitored by you or your business management. Direct labor expense is an important variable in determining the ultimate profit of the company. Labor percentage is also a direct measurement of the efficiency of the workers and performance of supervision. The variables in labor durations are the hardest expense to estimate accurately.

Direct labor burden. This includes all payroll taxes, plus payments for insurance and employees' benefits, which are associated with the labor payroll. If the workers belong to a union, the direct labor burden will include union benefit assessments and, in some cases, association fees.

Other direct costs. This includes all items other than those listed above, which are directly chargeable to individual jobs. For example, permits, bonds, insurance, and equipment rentals would be included here.

General and administrative expenses. On the income statement, this figure comprises all items of expense of a general nature which cannot be specifically attributed to individual or specific projects. These expenses are summarized for the year on a separate schedule, which supplements the income statement. The extent to which individual expenses will be listed separately or combined with others will vary among accountants, according to their methods of bookkeeping.

Pretax income. To get this figure, total operations expenses are subtracted from sales revenue. This figure is called income before provision for federal income tax on those corporations which are subject to federal taxation. Certain corporations, as well as sole ownerships and partnerships, do not pay, taxes on income; the income is reported as personal income on the owners personal tax returns. This is another example of how the business organization form you initially choose will influence the content of your financial statements.

Net profit for year. This figure is often called net income and represents the sum of all revenues minus all expenses including taxes, if applicable. Net profit or net income is commonly referred to as the "bottom line."

The Importance of Ratios

The bank loan officer who is to recommend the establishment of a line of credit (short-term loan) would be interested in the current and quick ratios, as the bank will expect you, the business owner, to repay the loan in the near term. The long-term creditors (banks or insurance companies) would be interested in the ratio of total liabilities to total assets. This measure indicates the relative proportions of the company's assets supplied by creditors and by owners. in the event that the company defaulted on its debts, this ratio indicates the degree of safety for the creditors. The bank loan officer will normally require balance sheets and income statements on the business for the current period, as well as for several prior years. You, the small business owner, are advised to have these reports prepared by a Certified Public Accountant (CPA) to ensure that the financial statements meet professional accounting standards.

Net sales to net working capital, net sales to net worth, and net income to total assets are all measures of the efficiency of the company's use of its resources. These measures are important indicators to management. Net sales to net working capital (low ratio) might be attributed either to an excess of working capital or to inadequate sales. You should examine each ratio. If the ratios reflect a weakness the manager then must analyze the problem area. For example, a more vigorous collection effort might be needed to reduce the size of accounts receivable. Keep in mind that the ratios are also interdependent. When receivables are reduced, the cash generated by collection efforts may be used to reduce long-term liabilities and improve the debt ratio. You as a small business owner must not focus your attention narrowly, or the resulting loss of control can lead to business failure. The broad perspective created by knowledge of business principles is essential for the success of your company.

Financial analysis and ratios. While the figures listed in the financial statements are meaningful and important when taken alone, they become even more valuable when compared to other information. For example, comparing any balance sheet item for the current year with the prior year immediately gives us added information. Did the figure increase? Decrease? If so, by how much in absolute dollars? How much in percentage terms?

The comparison of financial relationships is often done in three ways. The first method is the comparison of current financial data with prior years. Very often businesses will compare three to five years of key items, such as revenue and net income. This type of comparison gives the reader an idea as to the trends over time. The second method is to compare the current financial data with that of other businesses within the same industry. The third method is ratio analysis. The relationship between any two figures within the current financial data is called a ratio. For example, if current assets are $100,000 and current liabilities are $50,000, the relationship of current assets to current liabilities ($100,000/$50,000) may be shown as 2 to 1, or 2:1.

Current ratio. One very important kind of information that can be readily determined from the balance sheet is the ability of the company to pay its debts when due, often called liquidity. The difference between the total current assets and the total current liabilities is called working capital. One way of looking at working capital is to say that it represents the amount of money that is free and clear if all current debts were paid off. A comfortable margin of working capital gives a company the ability to meet its debt obligations and take advantage of opportunities. What is a comfortable amount of working capital for you to provide this margin? The current ratio is used to provide additional helpful information. To calculate the current ratio, divide current assets by current liabilities. For example, if your current assets equalled $250,000, and your current liabilities equalled $180,000, your current ratio would be:

$$\frac{\text{Current assets}}{\text{Current liabilities}} = \frac{\$250{,}000}{\$180{,}000} = 1.39 \quad \text{or} \quad 1.4$$

Therefore, for each $1.00 of current liabilities there is $1.39 in current assets to back it up, leaving $ .39 of every $1.00 for net working capital, which is money to work with. Your small business will flourish with proper use of these ratios.

Quick ratio. Your business' quick ratio is another, more conservative way of testing the adequacy of its current liquidity. Instead of using current assets, quick assets are substituted because these are quickly converted into cash. One simple method of determining the quick assets is total current assets minus inventories. Using the example above, if your inventories equal $110,000, the formula would be:

Current assets............	$250,000
Inventories..............	- $110,000
Quick assets..........	$140,000

Quick Ratio:

$$\frac{\text{Quick assets}}{\text{Current liabilities}} = \frac{\$140{,}000}{\$180{,}000} = .78$$

Thus we have $.78 of assets that may be readily converted into cash for each $1.00 of current liabilities that will require cash payments shortly.

Net sales to net working capital. This ratio shows how many dollars of sales the business makes for each dollar of working capital. A low ratio might indicate that working capital is not being efficiently used to generate sales. A high ratio might signal that there is not sufficient working capital to maintain sales. To calculate this ratio, divide net sales by net working capital. If your net sales equal $1,250,000 and your net working capital equals $70,000, the formula would be:

$$\frac{\text{Net Sales}}{\text{New working capital}} = \frac{\$1{,}250{,}000}{\$\ \ 70{,}000} = 1.79$$

Net sales to net worth. The net worth of a business can be derived by subtracting total liabilities from total assets. Thus, net worth is equal to net assets, or as in a corporation, the total stockholders' equity. The ratio of net sales to net worth is a measure of efficiency in the use of net assets. For example, if your net sales equal $1,250,000 and your net worth equals $120,000, the formula would be:

$$\frac{\text{Net sales}}{\text{Net worth}} = \frac{\$1,250,000}{\$\ 120,000} = 10.4$$

Total liabilities to total assets. This ratio compares the amount invested in the business by creditors with that invested by stockholders or owners. The higher this ratio, the higher the claim by creditors upon the assets of the company. This may be an indication that the business is extending its debt beyond its ability to repay. For example, if your total liabilities equal $280,000 and your total assets equal $400,000, this ratio would be:

$$\frac{\text{Total liabilities}}{\text{Total assets}} = \frac{\$280,000}{\$400,000} = .70\text{, or }70\%$$

Net income to total assets. This ratio is a measurement of the stockholders' rate of return on their investment and is typically given as a percentage. The ratio is achieved by dividing the company's net income by the total assets. If your net income equals $75,000 and your total assets equal $400,000, the formula would read:

$$\frac{\text{Net income}}{\text{Total assets}} = \frac{\$75,000}{\$400,000} = .1875\text{, or }19\%$$

Capitalization. This refers to the total financial resources made available to the owner. These financial resources are used to acquire the physical assets necessary to conduct the business. The more obvious requirements need to finance the physical resources used daily in the business. The need for additional financing to cover office expenses, licenses, payroll expenses, bonding, rentals, etc., because of the difference between when you pay and when you get paid, is critical for the business. A lack of enough working capital and insufficient cash liquidity generally result in business failure.

Billing retention. The amount of working capital required depends on the type of contracting business. The progress payments required under contracts for custom building and remodeling may be used to meet payroll expenses and material costs. The typical contract provides for three or more progress payments. Ninety percent of this comes during the construction phase and the final ten percent is held as retention. Smaller projects may require one-third upon starting the job, one-third at completion, and one-third at the end of the lien period. A contractor cannot collect more than the percentage already completed. The contractor must be aware that these excepted differences in the timing of expenditures and receipts could limit his capacity to finance the business, particularly if the company commits itself to new jobs before final payments on completed jobs have been received.

Significance of retention. This must not be underestimated. Retention usually exceeds profits and, therefore, represents a claim on working capital. Retention payments to the individual contractor may be held up through no fault of the contractor. The total project may have to be accepted before retentions are released. The problem is even greater for the subcontractor who completes his phase early in the project and must contend with a long waiting period to get retention. Since these funds are not available for uses elsewhere in the business, contractors must often finance their costs by borrowing. Speculative builders require larger amounts of capital than custom builders. This would imply that speculative builders must provide substantial financial resources to qualify for loan commitments. Too rapid growth can overextend the company to the extent that its solvency may be jeopardized. Financial planning helps avoid cash shortages.

Sources of financing. The new business owner typically lacks the required financial resources. Two types of external financing are available. Equity funds are supplied by investors who acquire some control of the business and a share of future profits, in return for their investments. Debt represents borrowed dollars, which require both the repayment of the original principal and periodic interest payments. The owner normally does not give up full control of the business.

Equity. Typical sources of these funds are acquaintances of the new business owner. Equity funds are sometimes available from private venture capital companies, small business investment corporations which are funded by the federal Small Business Administration, and minority enterprise small business investment companies. These sources are usually restricted to businesses with proven track records in a growing industry.

Long-term debt. Banks usually limit long-term borrowing for business ventures to 50% of the capitalization (debt ratio). Then, such funds will be made available only to business owners with good credit ratings, technical knowledge, and verifiable capacity for repayment as evidenced by financial planning, in the form of projected balance sheets, income statements, and cash budgets.

The Small Business Administration can make direct loans, but prefers to provide funds by guaranteeing up to 90% of a bank loan. The SBA will not review a direct loan application unless your firm has been refused by at least one bank; or two banks if yours city's population exceeds 200,000. The SBA prefers that an equal-dollar amount be put up by the owner for loans under either the guarantee program or the direct loan program.

Short-term debt. The operating business may want to establish a line of credit with a bank to meet some of its short-term needs for capital. This is called an "open account" credit line. Commercial banks favor short-term, self-liquidating loans of the type used to provide a line of credit to fund the extra work for which the money was needed. Such loans are typically repaid when the work is finished.

Profit. When all the costs have been totalled, the business owner must add profit. Profit is essential to the continuation of the business, and represents the return for bearing risk. A new owner's instinct is to price his service cheaper than the competition's to get started, but the owner will seriously jeopardize the business by omitting an allowance for profit. Insufficient profits threaten both capitalization (losses reduce retained earnings) and cash flow (payments to vendor vendors may exceed receipts) to the extent that bankruptcy and/or dissolution can be the result.

New Technology

IBM has developed a new laptop computer-radio with a fax and data modem that will instantly send your commands from a remote area, without phone lines, to the mother computer back in the office. They say that right around the corner, they'll be able to offer a VCM attachment for it -- that is a Video Camera Monitor link-up with the folks back in the office. This gives a new meaning to "face the fax." The next generation of technology is already upon us. If you are to be effective as a modern project scheduler, you must be on top of the technology coming down the pike.

Many new software packages specializing in estimating, cost control, and contracts administration will now interface with ***FastPro*** and other CPM scheduling programs. These new-generation programs offer advanced features such as remote-node and remote-control support. In a remote-node environment, you can dial your office network and your remote PC acts as if it were connected to the network. Remote-node connections exchange "packets" of network data together with your project data, and these packets often slow the transfer of data. Data transfer via remote control is typically faster than via remote node. In a remote-control environment, your PC (the remote PC) actually takes control of another PC (the host PC). You exchange only keystrokes and screen updates across the telephone line via modem, and the host PC processes all the data. This greatly reduces the amount of data transmitted across the line, because the remote PC does not have to transfer data processing commands. The power and configuration needed to drive the program resides in the host computer.

In a remote-access environment, multiple users can dial into your system simultaneously. Specifically dedicated software installed on both the host PC and on all remote PCs enables telecommunication. When you dial into the host, you can automatically choose the application you want to execute. Once in the application, you can update your project schedule data as though you have the program installed on your local drive. Some of these new remote access packages offer both remote-node and remote-control capability; however, remote control is preferred because of its faster data transference.

Remote-control programming is installed on the host PC, which becomes the application server to manage multiple users dialing in remotely. Because these new programs run on top of a customized version of OS/2, a 10-user-or-more version on the application server requires a 486 EISA PC with 16 MB of memory, plus an additional 3 MB of memory for each user. The application server can also function as a node if your company uses a Novell network. However, if you want dial-up capability for more than two simultaneous users, a multiport asynchronous adapter is required. As a remote user, you can access the application server by running the remote link on the source (remote) PC to the target (host) PC.

The host PC performs all the processing on the server, and the remote PC receives the screen updates. If performance is important, focus on CPU speed on the application server. Modem speed is also important if you need to improve performance. If you are dialing into a modem that is slower than the modem you are using, your modem will automatically drop down to the lower transfer rate. For maximum performance in these systems, the modems at both ends should be the fastest you can afford.

Chapter

17

Scheduling Contingencies

Scheduling Contingencies

Scheduling controversies often arise in the construction process, not only fromdisagreement over poor planning in the primary stages but also from operations among construction methods or sequences of activities and the project team and activity subcontractors in the secondary production stages. Often these disputes occur where a project is constructed under multiple prime contracts, such as those commonly experienced in fast-tracking, phased construction, or in a professional construction management contract. The absence of any contractual relationship between the individual prime contractors and the architect has been the main cause of numerous legal disputes.

A cursory routine approval of contractor-prepared schedules can be the most costly mistake an owner can make. In the absence of a thorough review by an experienced CPM project scheduler, contractually required schedules can become the primary source of documentation from successful contractor claims. The contractor can simply provide as evidence in court its prebid, preconstruction, and progress schedules, and then compare them with the as-built or adjusted schedules. The comparison can be very graphic, especially to those who have not had the training, as you have, to know how CPM works and how it can be subverted. If the owner approved the contractor's schedule, the owner can be left in an extremely weak legal position. This kind of comparison, to laypeople, is simple and straightforward. It appears to be the truth, pure and simple, and will have a significant impact on a panel of arbitrators or a jury.

But as we've already learned, the truth is rarely pure and never simple. CPM network schedules reflect those who produce them. Project schedulers who understand the nomenclature and advantageous uses of CPM also realize how it can be abused. The only thing between the owner and bankruptcy in scenarios like this are adequate records and a bulletproof CPM project schedule. So the professional project scheduler knows what scheduling contingencies are, and how to plan for them in the best interests of the client.

Claims against the owner in a CPM network schedule are, far and away, most often generated from changes in the schedule during production operation phases.

Scheduling Changes

Any scheduling change can have an impact on total project operations, and can even push back the contractual completion date. When scheduling changes are implemented as a result of constraints created by the owner or architect, the basis for a contractor claim may exist. Whenever the owner or architect issues any change order, field order, or work directive, or whenever a constructive change condition exists in any of the following areas, the contractor will begin to consider the impact of such action by the owner or architect upon its project costs and profitability.

These are the typical schedule contingencies that need advance alternative planning. They can lead to schedule changes that can later develop into owner-or architect-caused delays:

- Requirements to deviate from the schedule
- Orders to expedite the work
- Job interference by the owner or architect
- Owner constraints on scheduling interfaces
- Late availability of owner-furnished materials
- Owner-imposed acceleration or deceleration
- Impractical or impossible milestones
- Extra work orders
- Contract change orders
- Schedule impacts
- Sequence of construction altered by owner or architect
- Changes in activity's start event
- Changes in activity's finish event
- Time extensions
- Utilization of scheduling float time
- Late schedule approvals

If a CPM schedule is not available during the production, any consistent scheduling mehtod that illustrates the dealy will help support the claim. Any change in the CPM schedule, imposed by either the owner or the architect, can become a basis for a claim by the prime contractor or activity subcontractor. The best way to prevent a claim is to utilize CPM scheduling techniques. This form of scheduling documentation offers the greatest protection. As the project progresses, time extensions may be granted by the owner, or by the architect with owner approval. These would result in changes in the project schedule and could result in revised interface dates or changed completion dates.

The contractor will probably keep a close watch on the status of the schedules that it submits to the owner or architect for approval. Departures or deviations in the contractor's schedule, from the schedule originally intended by the owner, may be considered as contract amendments after approval by the owner.

If a contractor submits a shorter schedule than originally called for under the contract, and it is approved by the owner or the architect, any liquidated damages provided under the contract may now be applied to the early finish date shown in the shortened schedule. There is no question that the only way for a contractor to make a profit on a project is to get in and get out in the shortest possible time, thus reducing the home office overhead and labor costs allocated to that job.

Constructive Changes

A constructive change is an informal act authorizing or directing a modification to the contract, caused by the owner or architect through an act or a failure to act. In contrast to the mutually recognized need for a change, certain acts or failures to act by the owner, that increase the contractor's costs and/or time of performance, may be considered grounds for a change order.

This is termed a *(constructive change.)* However, it must be claimed in writing by the contractor within the time specified in the contract documents. Otherwise, the contractor may waive its rights to collect. Types of constructive changes include:

- Defective plans and specifications
- Architect or engineer interpretation of documents
- Higher standard of performance than specified
- Improper inspection and rejection of work
- Change in the method of performance by owner
- Change in the production sequence by owner
- Owner nondisclosure of pertinent facts
- Impracticability or impossibility of performance

Changed Conditions

Sometimes referred to as "unforeseen conditions" or "differing site conditions." the term *changed conditions* is typically used in all federal contracts. There is also a growing trend by many public agencies and a few private owners to adopt similar wording in their contracts. Failure of an owner to provide payment for changed conditions places the contractor in a difficult position. If the owner takes a hard-line position on this issue, the contractor may find it necessary to seek relief from the court, a process that is both lengthy and costly to both parties.

The federal policy is to make adjustments in time and/or price where unknown subsurface or latent conditions at the project site are encountered by the contractor. The purpose is to have the owner accept certain risks and thus reduce the large contingency amounts in bids to cover such unknown conditions. The federal government and many local agencies include provisions in their construction contracts that will grant a price increase and/or time extension to a contractor who has encountered subsurface or latent conditions.

Under the legal conditions for changed conditions, an existing underground pipeline that was either not shown on the plans at all or was incorrectly located on the contract drawings would qualify as a changed condition. Unusually severe weather conditions for the time of year and location of the project may also qualify. The discovery of expansive clays in the excavation areas, if not accounted for on the geological site survey and not detected in prior soil investigations, also qualify as changed conditions. Severe rains or similar weather that prevent work from being done, or which in any way delay the project, may not always be excusable delays, and in some cases have been ruled by the court to be excusable only and not compensable.

Schedule Acceleration

Acceleration of the work is usually the result of an attempt by the contractor to apply whatever means and take whatever measures are necessary to complete the work sooner than would normally be expected for a given project under stated contract conditions, or an attempt by the contractor to take extra measures to make up for delays, whatever the cause is, by utilizing whatever means are at the contractor's disposal to accomplish such an objective.

There are two recognized types of schedule acceleration:

1. Directed acceleration
2. Constructive acceleration

A directed acceleration occurs when the owner or architect, at the owner's direction and authority, orders a contractor to speed up the work. The owner is definitely in this legal position if it advanced the contractor's finish date.

The necessary elements of a constructive acceleration claim have been outlined by the U.S. District Court as follows:

"Constructive acceleration is present when
1. The contractor encountered an excusable delay entitling it to a time extension;
2. The contractor requested an extension of duration time;
3. The request was refused;
4. The contractor was ordered to accelerate its work, that is, to finish the project as scheduled despite the excused delays; and
5. The contractor actually accelerated the work."

In considering the liability of acceleration, the court's first consideration must be whether the so-called "acceleration" is the result of an order from the owner, or its agent, to a contractor who is behind schedule to get back onto schedule, or an order from the owner, or its agent, to the contractor requiring that the contractor, either directly or constructively, to complete the work prior to the scheduled completion date.

If the contractor is behind schedule and it is the desire of the owner, or its agent, to require the contractor to get back on schedule, it is sometimes necessary to direct the contractor to take whatever means are necessary to assure completion by the originally scheduled completion date. Herein lies an administrative risk for the project scheduler and contract administrator. If an order is issued to the contractor directing it to accelerate the work so as to catch up or make up for delays or lost time, the risk is that the contractor will perform as directed, then submit a claim for directed acceleration, arguing that although it appeared from the original schedule that it would not complete on time there was no risk of failing to complete on time at all.

The contractor might state that in the absence of the acceleration order it would have finished on time anyway. The proper handling of a situation such as this is not to issue an acceleration order (even though acceleration is justified or needed), but to send a letter to the contractor calling attention to the fact that completion by the scheduled date will be required, and that "according to the schedule it appears that the contractor will be unable to complete the activity by the completion date indicated in the contract. A revised schedule shows how the contractor plans to complete the activity by the scheduled date." This leaves the means and methods to the contractor, as well as responsibility.

Often the problem starts in the architect's design office. Project completion dates or time available to complete is frequently established by an architect or design professional who has had little field experience, and thus is hardly qualified to properly establish the real-world time duration needed to complete all the activities of the project. Typically you will see examples of this in activity durations that are too long or too short.

If insufficient time is allotted to the completion of a project, all the bidders will be forced to bid the job as accelerated work, thus increasing the costs materially. One indication that this may be happening comes when all the bids are in and, although they are grouped close together (a good sign of competent documents), they are all considerably in excess of the architect's estimated project cost. This can be evidence that all of the bidders are bidding on the job as accelerated work. This will increase the cost of construction above that for the same job completed within a normal CPM schedule.

Another risk surfaces here as well. After figuring a bid for a job as being all accelerated work, a bidder may look in the contract documents to determine how much liquidated damages are being assessed in case of a completion delay. Upon finding daily liquidated to be assessed, the bidder refigures the bid to include that amount.

In the foregoing example, let us assume that we plan to build a project that on a normal CPM schedule should take 16 months to complete, but the contract documents actually indicate the completion must be within 14 months, including two months' acceleration of work. Then the bidder will most certainly check the amount of liquidated damages called for in the contract documents. Let us presume the amount of liquidated damages was only $500 per day. An enterprising bidder will now go back to its 14-month bid and refigure the costs for completion in 16 months instead of the required 14 months. Then the bidder will determine that the difference in time that was allowed by the owner for completion versus the time that a job such as that should normally take was two months, or 44 working days. The bidder now multiplies 44 times $500 per day and adds the amount of liquidated damages to its total bid for doing the work in 16 months, and submit it as its bid price.

The problem here is that the cost of liquidated damages, at $500 per day, is considerably less than the daily cost of acceleration to the contractor; thus, the bidder submits a bid claiming that it will complete on time, when in fact it planned from the very beginning to complete the work two months late and pay liquidated damages costs. The owner and architect turn out to be the only persons who were unaware that the project was destined to be late before it was even started. Generally speaking, any job that has a resident project manager operating out of an on-site field office can justify in excess of $1000 per day of liquidated damages without difficulty. Typically in commercial construction in California, this figure starts at $2000 per day and goes upward.

Only then will liquidated damages serve as a deterrent to the contractor for finishing late.

A related condition called "deceleration" can also be experienced on a project's schedule. This occurs when a contractor is directed in writing or constructively to slow down its job progress. Many of the same considerations that apply to acceleration also apply to deceleration.

In preparing for a claim for acceleration or deceleration, it should be borne in mind that the costs to the contractor for going into premium time, such as working an extended work week, cannot be computed simply as including the added hourly costs multiplied by the additional hours. Studies have shown that, as the work week is extended, there is an accompanying drop in worker productivity. Furthermore, as the extended overtime is continued, the productivity rate continues to drop.

Project Schedule Productivity Losses

Scheduled overtime is not often seen on competitively bid lump-sum contracts, as most contractors are well aware of the negative effects of overtime on cost and productivity. Simple arithmetic shows that premium pay for double time or time-and-one-half makes overtime work much more expensive.

However, people who insist on overtime seldom realize that other costs associated with overtime may be even more significant than premium pay. Premium pay affects only overtime hours, but continuing of scheduled overtime drastically affects costs of all hours. All available research indicates a serious inverse ratio between the amount and duration of scheduled construction overtime and the labor productivity achieved during both regular and overtime hours.

Studies have shown that, in the first few weeks of scheduled overtime, total productivity per person is normally greater than in a standard forty hour work week, but not as high as it should be considering the number of additional work hours. After seven to nine consecutive fifty or sixty-hour work weeks, the total weekly productivity is likely to be no more than that attainable by the same workforce in a standard forty hour work week. Productivity will continue to diminish as the overtime schedule continues. After another eight weeks or so of scheduled overtime, the substandard productivity of later weeks can be expected to cancel out the costly gains made in the early weeks of the project's overtime schedule. This means that the total work accomplished during the entire period over which weekly overtime was worked will be no greater, or possibly even less, than if no overtime had been worked at all.

When the loss of productivity is combined with the higher wage cost (including premium pay), productive value per wage dollar paid, after several weeks of scheduled overtime, drops to less than seventy-five percent for five ten-hour hour days, less than sixty-two percent for six ten hour days, and less than forty percent for seven twelve hour days. When an overtime schedule is discontinued, it has been found that there was a dramatic jump in productivity per hour after return to a forty-hour work week.

Construction delay claims involving acceleration of the work usually include claims for loss of productivity, which often exceed all other claimed amounts. The following breakdown of a claim on a public works job illustrates the relative magnitude of the claim for loss of productivity, as compared with the other issues shown:

1. Extended project overhead	$ 1,019,099
2. Unabsorbed home office overhead	227,620
3. Labor escalation	142,430
4. material escalation	148,329
5. Labor loss of productivity	2,442,409
6. Subcontractor claims	920,407
SUBTOTAL	$ 4,900,284
7. Profit and overhead on items 1 to 5	1,298,575
8. Unresolved changes	157,993
9. Interest on money	1,073,897
10. Additional bond premium	11,844
TOTAL CLAIM	$ 7,442,593

Suspension of the Project Schedule

The work on a project can usually be suspended by the owner for any of several reasons. In each case, the owner or its agent should keep detailed cost isolation records of all activities affected by the suspension. It should be kept in mind that suspension of the work for any amount of time, such that the completion date is extended, may impact the contractors' costs through unabsorbed home office overhead and the real possibility of missing other projects due to the delay.

The contractors may also claim the effect upon their organizations of the costs related to dismantling operations, mobilization or demobilization, direct costs, settlement expenses, escalation costs, prior commitments, post-termination continuing costs, unabsorbed overhead, unexpired leases, severance pay, implied agreements, restoration of work, utility cutoffs, inventory, replacement costs, and all other allocable costs. On suspensions of the work, be certain that all such orders are in writing and that a careful detailed, record is kept of its total effect on each contractor's time and costs.

Change Order Pricing

One of the most common causes of contractor claims occurs during attempts to price change orders. Typically, owner change orders contain a waiver clause that requires the contractor to guarantee that the price and time named in each individual change order represents the total cost to the owner for that change, and the contractor waives any rights to impact costs.

This, unfortunately, leaves the contractor only one recourse -- the claims process. The claims courts have traditionally held that an owner may not force a contractor to "forward price" its impact costs when pricing an owner-directed change. The court has written: "While it might be good contract administration on the part of the owner to attempt to resolve all matters relating to a contract modification (change order) during the negotiation of the modification, use of a clause which imposes an obligation on the contractor to submit a price breakdown required to cover all work involved in the modification cannot be used to deprive the contractor from its right to file claims."

Project Schedule Conflicts

This is an often misunderstood area in contractor claims against the project. However, the probability of recovery by the contractor as the direct result of such conflicts is good, insofar as the settlement is limited to the cost difference between the project cost as the plans and specifications are interpreted by the owner or architect and the contractor.

Public works contracts are called *contracts of adhesion*, which is a term applied to contract documents that are drawn by one party and offered to the other party on a take-it-or-leave-it basis, where there can be no discussion of terms nor contract modification by the other party. The contractor, however, does have one advantage. In case of ambiguity, the court will interpret the contract in the contractor's favor. This does not relieve the contractor from the obligation of building the work in accordance with the interpretation of the architect or design professional, but only assures that the contractor be paid.

Frequently, the contractor will find that outdated standards are specified, or that products are named that no longer exist. Often, the specification will contain references stating that, wherever codes or commercial standards are specified, the contractor is obligated to use the latest issue of that standard existing at the time the project went to bid. Unfortunately, in many cases the designers considered the fact that the design was based upon an old standard that existed in their files, or a standard that was current during the design phase, but which later may have been updated by the sponsoring agency without the designer's knowledge. Occasionally, serious difficulties arise from such practices, and the contractor has the legal right to assess the project cost difference resulting from the error. Perfect specifications are hard to find, so the contractor must make a reasonable interpretation at the time of bidding the job and will then be in a good position for recovery if a variance exists. The contractor's interpretation must be based on what a reasonable person would interpret the documents to mean; then the contractor will have the court on its side.

A contractor should never attempt to construct any questionable area, or change, without first submitting to the owner or architect a request for clarification, or notifying them of an error. This is known as an RFI (request for information), or an RFC (request for clarification). In most contract forms, failure to do this may bar full recovery of contractor costs.

Other problems that often come up in project scheduling are directly related to the conduct of the owner's representatives on the job, such as the project manager or site supervisor issuing changes in the work that are of such magnitude as to constitute a cardinal change which creats a breach of contract. Or, on a lesser scale but indirectly related, where one activity subcontractor may negligently delay another, which may result in the owner's seeking recovery from the negligent contractor to pay the contractor that was harmed.

Some of the types of scheduling problems that fall into this category include the following:

- Damage to work by other prime contractors in fast track scheduling
- Breach of contract
- Cardinal changes
- Beneficial use of the entire project before completion
- Work beyond contract scope
- Partial utilization of the project before completion
- Owner nondisclosure of site-related information
- Owner's failure to make payments when due

Scheduling techniques, such as CPM or other network scheduling systems and their associated computer-generated printout sorts, have also compounded problems between the project team and the work contractors. Often the owner is swamped with stacks of computer printouts that he either does not want or does not know how to read. Yet somewhere in that stack of paper may well be the key defense against claims being made against the project and the owner. The question of scheduling methods of operation, and the impact of deviating from the anticipated schedule, requires a careful analysis to determine the reasonableness of the originally anticipated schedule in conjunction with the planned methods of operation.

The analysis of every schedule-related problem is unique, and every such analysis should include a review of the anticipated sequence and schedule together with a review of the actual progress of the work. The review must include an analysis of any delays and impacts caused by all parties to the progress of the project.

It is in the presentation of the sorts, containing cost data and their supporting documentation, that most construction claims are the weakest. A carefully detailed reconstruction of all construction activities and their related costs must be made and presented so that they can be clearly identified for use in negotiating affirmatively or defensively. Construction-related organizations are reported to be poor in their cost recordkeeping systems (or lack of systems), and they traditionally lack ability to assign costs to possible claims areas.

An accounting firm is not the answer either, because often those cost items relating to potential claims are not identified by an accounting firm. Most accounting firms are not familiar with the cost-control procedures for construction contracts. A project scheduler who realizes this, and takes steps to structure and run cost-tracking and audit trail sorts, will provide his or her clients with the best form of protection against claims. Remember, records are the owner's first line of defense in construction litigation documentation; if not by the accountant, then by the project team. The responsibility may get assigned to the project scheduler for computer recordation and tracking.

The accounting operations must cover all the necessary payroll functions, and the following employment laws must have tracking.

- All wages earned by any person working for an activity subcontractor for hourly wages on the project are due and payable at least twice during each calendar month. Labor performed between the 1st and 15th days shall be paid for between the 16th and 26th day of the month. Work performed on the project between the 16th and the last day of the month shall be paid between the 1st and 10th day of the following month. However, the salaries of administrative or salaried management employees may be paid monthly.

- In a case of dispute over wages, the activity subcontractor shall pay all wages conceded by him to be due, leaving the employee the right to dispute the remaining amount.
- Every activity subcontractor employer shall keep posted conspicuously at its place of work a notice specifying the regular pay days and the time and place of payment.
- In the event of a labor strike, the unpaid wages earned by the striking employees shall become due and payable on the next regular payday.
- No activity subcontractor shall issue in payment of wages due any check check or draft unless it is negotiable and payable in cash upon demand. No activity subcontractor shall issue script, coupons or other things redeemable in merchandise as payment for wages.
- An activity subcontractor may not withhold from any employee any part of the wages unless required by the government.
- No activity subcontractor shall charge a prospective employee for any pre-employment medical or physical exam or charge an employee of the same.
- Every activity subcontractor shall, semi-monthly, at the time of each payment of wages, furnish each employee, either with a detached part of the check paying the employee's wages, or separately, and an itemized statement in writing showing:

  - Gross wages earned
  - All deductions, including SDI, FICA, State & Federal taxes
  - Net wages earned
  - Dates of work period
  - Name of employee or Social Security number
  - Name and address of employer

- Every activity subcontractor who pays wages in cash shall, semi-monthly or at the time of each payment of wages, furnish each employee an itemized statement in writing, showing all deductions, dates of work period, name of employee and Social Security number, and the name and address of the activity subcontractor.

- Whenever an activity subcontractor has agreed with any employee to make payments to a health or welfare fund, pension fund or vacation fund, it shall be unlawful for such activity subcontractor to willfully or with intent to defraud or fail to make the payments required by the terms of any such agreement.

- No activity subcontractor shall demand or require an applicant for employment or any employee to submit to or take a polygraph or lie detector test as a condition of employment or continued employment.

- No activity subcontractor shall require any prospective employee or any employee to disclose any arrest record which did not result in conviction.
- Eight hours of labor constitutes a day's work, unless stipulated expressly by the parties to a contract. No employee shall be required to work more than eight hours per day or more than forty hours a week unless time and one-half is paid.
- Each person employed is entitled to one day's rest in seven. Violation of the law is considered a misdemeanor.
- Labor laws usually apply to those hours exceed thirty hours a week. Employment that does not exceed thirty hours in one week or six hours in one day is considered part-time and is not protected by the same laws.
- No activity subcontractor shall force any employee to join or not join a labor union.
- An activity subcontractor who seeks to replace employees on strike or not working due to a lockout shall be permitted to advertise for replacements if such advertisement plainly mentions that a labor dispute is in progress.
- A strike means an act of more than 50% of the employees to lawfully refuse to perform work.
- A lockout means any refusal by an activity subcontractor to permit any group of five or more to work as a result of a dispute over wages, hours, or condition of employment.
- A jurisdictional strike means a concerted refusal to perform work for an employer arising out of a controversy between two or more labor unions as to which of the unions has or should have the right to bargain collectively with an employer. This type of strike is illegal in the State of California.
- The Labor Commissioner may issue a citation to any activity subcontractor disobeying the labor laws. The activity subcontractor has, upon receipt of the citation, ten days to request a hearing of contestment.
- A collective bargaining agreement is an agreement between management and labor to which they have collectively agreed following a period of bargaining. These agreements are enforceable in California courts.
- Where a collective bargaining agreement contains a successor clause, such an agreement shall be binding upon any successor activity subcontractor who purchases the contracting activity subcontractor's business until expiration of the agreement or three years, whichever is less.
- Professional strikebreaker means any person who repeatedly offers himself for hire to two or more employers during a labor dispute for the purpose of replacing an employee involved in a strike or lockout. Such strikebreakers are illegal in the State of California.

- A minor under sixteen years of age is forbidden to:

    - Work on machines, power tools, saws, etc.
    - Work on scaffolds.
    - Work on excavation work.
    - Drive vehicles
    - Work with dangerous or poisonous acids
    - Work in tunnels or other types of excavation.

- No discrimination shall be made by any activity subcontractor in the employment of persons because of race, religion, color, national origin, ancestry, physical handicap, medical condition, marital status, or sex. Nothing shall prohibit an activity subcontractor from refusing to hire or from discharging a physically handicapped person unable to perform duties or whose performance would endanger his health or the health or safety of others.
- It is unlawful for an activity subcontractor to refuse to hire or employ, discharge, dismiss, suspend, or demote any individual over the age of forty on the grounds of age unless the person fails to meet the bona fide requirements for the job. This shall not limit the right of an activity subcontractor to select or refer the better qualified person from among all applicants for the job.
- Activity subcontractors shall keep all payroll records, showing the hours worked daily, wages paid, etc., for at least two years.
- No activity subcontractor shall knowingly employ any alien who is not entitled to lawful residence in the U.S. if such employment would have an adverse effect on lawful resident workers.
- No deduction from the wages of an employee on account of his coming late to work shall be made by the activity subcontractor in excess of the proportionate wage which would have been earned. For a loss of time of less than thirty minutes, one-half hour may be deducted.
- Every activity subcontractor having one or more employees, whether temporary or part-time, is subject to the Social Security laws of the Internal Revenue Service.
- Each activity subcontractor must secure an employer's identification number from the Internal Revenue Service or from the Social Security Administration.
- Each activity subcontractor must ascertain each employee's correct Social Security account number and copy it directly from the employee's Social Security number card when he or she starts work.
- Every activity subcontractor who has one or more employees in his employ and pays wages in excess of $100 or more during a calendar quarter becomes an employer within the meaning of state tax laws, is responsible for

Unemployment Insurance Code regulations and is subject to its provisions. Everyone who becomes an employer is required to register within fifteen days with the state's Employment Development Department.

Chapter
18
Conclusion

PostScript

A project schedule is the contractual network diagram of the project's planned activities, their sequence determined by job logic, the contractual time in working days required for completion (activity duration), and the conditions necessary for their completion (contract specifications). It is also a contract document linking the lender, developer, prime contractor, and subcontractors. It serves, within the contract specifications, to advise the lender and developer of any unsatisfactory progress in any activity's production, and as a reminder to the prime contractor and activities subcontractors of the jobs they must accomplish within their contractual time frame. It is accepted industrywide that use of CPM network scheduling in modern construction projects is the difference between success or failure. CPM has proven itself to have these three fundamental advantages in project scheduling:

1. CPM scheduling provides instantaneous monitoring of the project's production.

2. CPM scheduling increases productivity and efficiency of the project's production.

3. CPM schedules are contract documents that stand up in court to prove or deny contractor delay claims.

We learned that, in the conception stage, project scheduling is the process of carefully considering all the activities of the project's production. Then comes the process of grouping them into phases, listing them out, then graphing them out. This is then followed by the planning stage, in which you determine the steps required to accomplish each activity, then lay out those steps in a logical production sequence. We called this *job logic*. Many things influence job logic, such as trades subcontractors' availability in your area, fabricated items delivery dates, or a trucking strike which could delay long-lead items delivery at the last moment. And any such delay could delay other activities and thus influence the job logic or activity sequence for your area, impacting your project schedule.

Next comes the schedule development phase in which you use the fundamentals of CPM project scheduling to coordinate the many tasks, activities, and phases necessary to bring that project into existence. The numbers are then crunched and the schedule is cooked, which means applying critical analysis to debug the system. Once the schedule is authorized by the lender and owner the project begins, and the project scheduler now starts doing cost-tracking and audit trail procedures. The professional project scheduler also begins doing quarterly trends analysis to see how the project schedule will do next quarter, to determine what changes are needed now.

The crucial part of successful project scheduling, however, is the methodology you use to complete these tasks. Profitable project scheduling requires knowledge of network scheduling and the use of modern computerized CPM programs. Integrated software products configured with dedicated program logic are the modern project scheduler's tools of the trade. Professional project scheduling always contains a dichotomy of conceptual planning then carried out in a real-world application. Seldom are the two conducive to easy production. Currently, the project scheduler is hired to develop the production schedule, overseeing the same, and reporting progress to the project manager, owner, project team, prime contractor, and subcontractors through specialized summary report sorts for each.

The project schedule must also interlink with the other major project considerations, such as design and long-lead items procurement. That is easy enough where, traditionally, the design is completed before the project goes to construction bidding. But in other production modes, such as fast-tracking, design should be thirty to fifty percent complete if a reliable operational project schedule is to be developed and executed. But when a scheduler gets this advance start on a project without a finished projects design, error is obviously probable if the works has not been done previously to provide accurate activities durations.

Even a detailed CPM schedule with a comprehensive work breakdown is not an adequate tool for controlling the daily progress of the activities in the field. Field scheduling is necessary to coordinate the reporting of data to the main scheduling system. On smaller construction projects, the production progress milestone schedule may be the only one used. Sequential completion of phases is simple enough on small jobs and usually will be adequate. On larger projects, however, the project schedule is also the basis for more detailed weekly work plans in the field for each major activity. In either case, the approved project schedule will, in turn, determine the type and detail of the field scheduling used by the owner and the prime contractor to set the production phases milestones and monitor the projected schedule progress versus the actual completed progress. Detailed field scheduling is necessary continually throughout the entire project to interlink the two for critical analysis and control.

Complete detailed field planning makes the most efficient use of field manpower on the priority list of tasks within those activities that are required to meet the CPM milestone dates. This detailed planning first concentrates on those items of work that are on the critical path, then secondary activities that are noncritical. The field scheduling personnel of the project team are responsible to the project manager for planning the construction activities each week for the following week. Field schedulers list out the work activities sequentially that are scheduled to be completed by the following week. This list of pending activities is then discussed in the weekly scheduling meeting with the project manager, project scheduler, prime contractor major subcontractors, the chief field engineer, and the site superintendent. These weekly meetings allow the project scheduler to confirm data and readjust if necessary. In any attempt to improve the production timeline, the project scheduler must still evaluate the schedule in relation to the contracting plan and the construction technology presently available.

Adios

Downsizing stories and job losses are all around us. If you haven't been affected you probably know someone who has. Even if you have a good job now you may be looking over your shoulder, wondering when your turn will come. While current studies reflect that most Americans are nervous about the future, in the construction industry the career angst hangs like a blanket. And it's making a profound change in our lives, whether we know it yet or not.

It's hard to make a statistical case that business downsizing has changed American life dramatically. The country is making a slow recovery from the latest recession, but some people are still driving luxury cars. Economists say downsizing has forced a lot of people out of jobs, but the worst is over. "The wave of mass layoffs at American corporations is subsiding," the Wall Street Journal recently proclaimed.

So if things are getting better, why don't we feel more bullish on America? Because, despite what the statistics say, life is tougher than it used to be for middle- and upper-middle-class workers. Good jobs are harder to come by and harder to keep. Financial security is increasingly elusive. For us baby boomers especially, downsizing has come as a shock. We've been incredibly lucky. Every generation before ours lived with the possibility that the plant might shut down, Dad would be killed in a war, the crops would fail, or the boss would get angry and fire the whole staff. The generation reared during the 1930s certainly knows what downsizing really looks like.

But 40-somethings and people now in their mid-50s grew up during a period of unbridled prosperity in America. We were spoiled. We thought there would always be money to be made, new frontiers to explore, big companies hiring us and lavishing us with raises, promotions, and eventually, gold watches. For those of us who expected to duplicate our parent's steady rise to affluence, downsizing has been the latest in a series of jarring jolts of reality. We were taught go to college, pursue one career, and do it the rest of our lives. Now current wisdom says you should be prepared for six career changes in your working life. Lifelong employment with any company isn't realistic anymore. You contribute what you can as long as you can, and then you retool.

The construction industry has been no exception. The new ways have changed everything, from design to production to project execution. But this change has also helped define and create demand for the professional project scheduler, in not only the construction industry but also in other businesses that run production systems in their operations. This wave of change will work to your benefit if you are sharp enough to climb aboard now. Take advantage of the modern scheduling techniques contained within this reference manual, along with the business combat tactics to protect your clients, and the business management functions in the consultant chapter. Downsizing leads to outsourcing. The companies that will build the projects of the future are looking for outside consultants like you.

It has been my intent to produce a comprehensive scheduling manual in conjunction with a parallel CPM software program. I hope I've been successful. This book focuses on the lucrative field of CPM construction project scheduling. It provides a comprehensive, up-to-date examination of project scheduling, comes with a ready-to-use software program, and does so at a more affordable price than any other professional guide. This book was written for both practicing project schedulers who want to advance their careers and for those planning to enter the field. The book and software cover the phases of modern construction scheduling, as well as all pertinent requirements, the documents and sorts, and related construction and contract law.

I've tried to inject humor and real-world applications to demystify CPM, because I've learned in my years as a teacher that one must put the student at ease. Only when one is relaxed does attention fully turn to the lesson at hand, and knowledge truly flow. My style of writing will undoubtedly be too loose for some looking for a spit-and-polish reference manual, but I come from the trades and I've learned from experience that people learn and retain more from working professionals than from isolated academics.

So I offer no apologies for my frontier literature style and hope that my editor, even though she's a button-down East Coast professional, will keep the saltier passages intact and allow my words to come home to you. It has been my great pleasure to serve as your author of this bookware product. When the knowledge contained herein becomes yours and begins to pay off in your career, it will be my great honor to have been your instructor.

Appendix

A

Project Production Abbreviations

Contract Documents, which include the project's schedule, are governed by AIA specifications. These AIA acronyms and abbreviations are typical and commonly used. You will note some of these abbreviations have periods while some don't. This is not a typo, this is the way they appear in use. Don't ask me why, because I don't know why. Neither does anyone else at the AIA headquarters office. It's just the way it is. Probably one of those things that tradition has passed on to us just to drive people nuts when they try to learn the system.

You will see many of these in production documents. Make sure you understand all acronyms and abbreviations that come to your production scheduling. You will note by examining these that there are subtleties that go with the periods, such as PCC stands for "point of compound curve" whereas P.C.C. stands for "Portland Cement Concrete". As always, pay attention to details in looking at abbreviations, especially in your long-lead items. Remember the Scheduler's Credo, "Check twice, act once."

A	Area
AASHO	American Assn. of State Highway Officials
A.B.	Anchor Bolt
A.B.	Aggregate Base
ABS	Acrylonitrile Butadiene Styrene (plastic sewer pipe)
AC	Alternating Current
A.C.	Asphalt Concrete
ACI	American Concrete Institute (concrete specs for concrete production)
ACP	Asbestos Cement Pipe
ACST	Acoustic
ACST PLAS	Acoustical Plastic
ACU	Air Conditioning Unit
AD	Area Drain
ADD	Addition
ADJ	Adjustable
AFF	Above Finish Floor (elevation)
AGG.	Aggregate
AH	Air Handling Unit
AIA	American Institute of Architects (contracts and design specs)

AISC	American Institute of Steel Construction (structural steel specs)
ALT	Alternate
AMT	Amount
ANGL	Angle
ANGL DF	Angle of Deflection (intersecting angle)
APP	Attactic prolypropylene
APPX	Approximate
APT	Apartment
ARCH	Architect(s) or Architectural
ASA	American Standards Association
ASHRAE	American Society of Heating, Refrigerating and Air Conditioning Engineers
ASME	American Society of Mechanical Engineers
ASPH	Asphalt
ASSEM	Assemble
ASSOC	Associate
ASTM	American Society of Testing Materials
AT	Acoustical Tile
AUTO	Automatic
AVG	Average
AWPA	American Wood Products Association (wood standards)
AWS	American Welding Society (steel specs standards)
AWWA	American Water Work Association (water standards)
Ba	Bay
BB	Bond Beam
BC	Begin Horizontal Curve
BF	Bottom of Footing
BHP	Boiler Horsepower
BITUM	Bituminous
BLDG	Building
BLKG	Blocking
BM	Bench Mark
BOT	Bottom
B.O.W.	Back Of Walk
BP	Blueprint (refer to plans)
BRG	Bearing (load bearing)
BRG PL	Bearing Plate
BRK	Brick
BRKT	Bracket(s)
BTR	Better
>BTR	Better Than
BTU	British Thermal Unit
BU	Bureau of Standards
BVC	Begin Vertical Curve
C	Thermal Conductive

C	One Hundred
CATV	Cable Television
CB	Catch Basin
CC	Center to Center
CEM	Cement
CF	Cubic Foot
CFM	Cubic Feet Per Minute
C&G	Curb & Gutter
CC	Center to Center
CL	Center Line
CLR.	Clear
CMU	Concrete Masonry Units (masonry blocks)
CMP	Corrugated Metal Pipe
CND	Conduit
CO	Company
CONC	Concrete
CONF	Conform
CONST.	Construct
CONT.	Contours or Continuous
CONTR	Contractor
CP	Concrete Pipe
CPM	Critical Path Method, Critical Path Management
CPM	Cycles Per Minute
CS	Cast Steel
CS	Commercial Standards (as in "up to")
CSF	100 Square Feet
CSI	Construction Specifications Institute (construction specs)
CTR	Contract (as in "per")
CU FT	Cubic Foot
CU IN	Cubic Inch
CU YD	Cubic Yard
D	Drain
D	Depth
DC	Direct Current
DEG	Degree
DEPT	Department
DET.	Detail
DF	Douglas Fir (construction grade lumber)
D.I.	Drain Inlet
DIA.	Diameter
DIAG	Diagram
DIP	Ductile Iron Pipe
DISC	Disconnect
DIV	Division
DL	Dead Load
DN	Down

DP	Duplicate
DPG	Damproofing
DSB	Double Strength B Grade Glass
D4S	Dressed & Matched on 4 Sides
DU	Double Strength
DWG.	Drawing D/W Driveway
DWV	Drain, Waste, Vent Piping
DX	Duplex
E	East
EA.	Each
E to E	End to End
EC	End Horizontal Curve
EL, ELEV.	Elevation
EJ	Expansion Joint
ENCL	Enclosure
ELEC	Electric
ELVR	Elevator
EMER	Emergency
EMT	Electrical Metallic Tubing (conduit)
ENGR	Engineer
ENT	Entrance
EP	Electrical Panel
EQ	Equal
EQUIP	Equipment
EST	Estimate
ETE	End To End
EWC	Electric Water Cooling
EXC	Excavate
EXCL	Exclude
EXH	Exhaust
EXIST	Existing
EXP	Expansion
EXR JT	Expansion Joint
EXP	Exposure
EXPO	Exposed
EXST	Existing
EXT	Exterior
EXTN	Extension
E.P.	Edge of Pavement
EQ	Equal
EVC	End Vertical Curve
EX.,EXIST	Existing
F	Fahrenheit
FA	Fire Alarm
F&I	Furnish & Install
FAB	Fabricate

FAO	Finish All Over
FB	Flat Bar
FBM	Foot Board Measure
FD	Floor Drain
FD	Found
FDN	Foundation (East Coast)
FE	Fire Extinguisher
FEC	Fire Extinguisher Cabinet
FED	Federal
FED SPEC	Federal Specification
FG	Finish Grade
FH	Fire Hose
FHC	Fire Hose Cabinet
FICA	Social Security Tax
FIG	Figure
F.L.	Flowline
FIN	Finish
FIN CEIL	Finish Ceiling (elevation)
FIN FL	Finish Floor (elevation)
FIX	Fixture(s)
FL	Flashing
FL	Floor
FLASH	Roof Flashing
FLG	Flooring (type)
FLNG	Flange
FLUOR	Fluorescent
FND	Foundation (West Coast)
FOB	Free On Board
FOC	Face Of Concrete
FOF	Face Of Finish
FOS	Face Of Studs
FP	Fireplace
FPM	Feet Per Minute
FPRF	Fireproof
FPS	Feet Per Second
FR	Frame
FRG	Furring
FRP	Fiberglass Reinforced Plastic
FS	Full Size
FT	Foot
FT^3	Cubic Feet
FTG	Footing
FTG	Fitting
FUT	Future (use)
FUTA	Federal Unemployment Tax

FWH	Frostproof
FX WDW	Fixed Window
G	Gas
GA	Gauge
GAL	Gallon
GALV	Galvanized
GB	Grade Beam
G.B.	Grade Break
GFI	Ground Fault Interruptor
GFCI	Ground Fault Circuit Interruptor
GI	Galvanized Iron
GL	Glazed (glass)
GND	Ground
GOVT	Government
GPH	Gallons Per Hour
GPM	Gallons Per Minute
GR	Grade
GR	Grate
GRAN	Granular
GRD	Ground
GRTG	Grating
GYP	Gypsum
GYP BD	Gypsum Board (sheetrock drywall)
HA	Hot Air
HB	Hose Bib
HC	Heating Coil
HD	Head
HDP	Head pressure
HDW	Hardware
HDWD	Hardwood
HEX	Hexagonal
HGT	Height
HM	Hollow Metal
HOR, HORIZ	Horizontal
HP	Horsepower
HR	Hour
HSE	House
HTR	Heater
HU	Humidifier
HV	Heating & Ventilation Unit(s)
HVAC	Heating, Ventilation & Air Conditioning
HW	Hot Water
HWH	Hot Water Heater
HWR	Hot Water Return
Hz	Hertz
I	I Beam

I	Iron
ID	Inside Diameter
IF	Inside Face
IMC	Intermediate Metal Conduit
IN	Inch
IN^3	Cubic Inch
INC	Incorporate(d)
INCAND	Incandescent
INCL	Include(d)
INCR	Increaser
INS	Insulate
INT	Interior
INV	Invert (flow line)
I.P.	Iron Pipe
INV.	Invert
J	Junction
JAN	Janitor
JB	Junction Box
JB	Jamb
JP	Joist & Plank
JST	Joist
J.B.	Junction Box
K	Kilo (1000)
KAL	Kalamein
KD	Kiln Dried
Km	Kilometer
KP	Kick Plate
KVA	Kilo Volt Amperes
KW	Kilowatt
L	Left
L	Lumen
L	Length
LAB	Laboratory
LAM	Laminated
LAT	Lateral
LAU	Laundry
LAV	Lavatory
LB	Light Beam
LB	Pound
LBR	Lumber
L.C.	Length of Curve
LDG	Landing
LEV	Level
LF	Linoleum Floor
L.F.	Linear Foot

LG	Long
LH	Left Hand
LHR	Left Hand Reversed
LIN FT	Linear Feet
LL	Live Load
LOA	Length Overall
L&PP	Light & Power Panel
LP	Low Point
LPG	Liquified Petroleum Gas
LR	Living Room
LS	Lump Sum (Contract)
LT	Light (Pane of Glass)
LTH	Lath
LVR	Louver
LW	Light Weight
M	Bending Moment (force)
M	Meter
M	Thousand
MAGN	Magnesium
MAS	Masonry
MATL	Material
MAV	Manual Air Vent
MAX	Maximum
MBF	Thousand Board Feet
MBtu	Thousand British Thermal Units
MBR	Master Bedroom
M&D	Matched & Dressed
MDO	Medium Density Overlaid
MDP	Main Distribution Panel
MECH	Mechanical
MED	Medium
MET	Metal
MFD	Manufactured
MFG	Manufacture
MH	Manhour
MH	Manhole
MI	Mile
MIN	Minute
MIN	Minimum
MISC	Miscellaneous
MIX	Mixture
MK	Mark
ML	Metal Lath
MLF	Thousand Linear Feet
MDG	Molding
mm	Millimeter

MN	Main
MN	Mean (average)
Mo	Month
MO	Motor Operated
MOD	Modular
MOR	Mortar
MS	Manual Starter
MSF	Thousand Square Feet
MTD	Mounted
N	North
NATL	National
NBS	National Bureau of Standards
NEMA	National Electrical Manufacturers Association
NF	Near Face
NFPA	National Fire Protection Association
NIC	Not In Contact
No	Number
NOM	Nominal
NOR	Normal
NRC	Noise Reduction Coefficient
NTS	Not To Scale
O to O	Out to Out
OA	Overall
OBS	Obscure
OC	On Center
O.C.	On Center
OCT	Octagonal
OD	Outside Diameter
OF	Outside Face
OFF	Office
OPNG	Opening
OPP	Opposite
OS	Outside
OUT	Outlet
OVHD	Overhead
OZ	Ounce
P	Plumb
PAR	Parallel
PART	Partition
PASS	Passageway
PAV	Paving
PCC	Point Of Compound Curve
P.C.C.	Portland Cement Concrete
PCR	Point of Curb Return
PCS	Pieces

PERF	Perforated
PERIM	Perimeter
PERP	Perpendicular
PG	Page
PH	Phase
PL	Pilot Light
PL	Plate
PL	Property Line
PLAS	Plaster
PLAT	Platform
PLBG	Plumbing
PLGL	Plate Glass
PL HT	Plate Height
PLMB	Plumb
PLY	Plywood
PNL	Panel
PPM	Parts Per Minute
PRCST	Precast
PR	Pair
PRC	Point of Reverse Curve
PREFAB	Prefabrication
PRESS	Pressure
PRI	Primary
PROP	Property
PROP	Proposed
PRV	Pressure Reducing (Regulating) Valve
PS	Pull Switch
PSI	Pounds Per Square Inch
PT	Point
P.T.	Pressure Treated
P.U.E.	Public Utility Easement
PVC	Polyvinyl Chloride (plastic water pipe)
P.V.I.	Point of Vertical Intersection
PVMT	Pavement
QT	Quarry Tile
QT	Quart
QTY	Quantity
R	Radius
R	Thermal Resistance
(R)	Radial
RA	Return Air
RAD	Radius
R.C.	Relative Compaction
RCP	Reinforced Concrete Pipe
RD	Round
RDM	Random

REC	Recessed
RECP	Receptacle
REF	Reference
REFL	Reflective
REFR	Refrigeration
REG	Register
REM	Remove
REINF	Reinforcing
REP	Repair
REQ	Required
RESIL	Resilient
RETG	Retaining
REV	Revision
RF	Roof
RFG	Roofing
RGH	Rough
RGH OPNG	Rough Opening
RH	Right Hand
RHR	Right Hand Reversed
RHW	Heat & Moisture Resistant Rubber
RIO	Rough-in Opening
RIV	Rivet
R/L	Random Lengths
R/W/L	Random Widths & Lengths
RM	Room
RO	Rough Opening
RPM	Revolutions Per Minute
RST	Rigid Steel Conduit
RT	Rubber Tile
RT	Right
RV	Relief Valve
RDW	Redwood
RWL	Rain Water Leader
s	Slope in Feet Per Foot
S	South
SA	Supply Air
SAD	Supply Air Defuser
SAF	Safety
SAN	Sanitary
SC	Scale
SC	Self Closing
SC	Solid Core
SCHED	Schedule
SCT	Sewer Clay Tile
SCUP	Scupper

SD	Storm Drain
SDMH	Storm Drain Manhole
SEC	Second
SEC	Secondary
SECT	Section
SER	Service
SEW	Sewer
SFCA	Square Feet of Form in contact with Concrete
SHLP	Shiplap
SHT.	Sheet
SHTG	Sheathing
SIM	Similar
SJ	Steel Joist
SM	Sheet Metal
SOV	Shut Off Valve
SPEC	Specification
Sq	Hundred Square Feet
SQ	Square
SQ FT	Square Foot
SQ YD	Square Yard
SR	Supply Register (air)
SS	Sanitary Sewer
SSB	Single Strength B Grade Glass
SSG	Single Strength Glass
SSMH	Sanitary Sewer Manhole
SST	Stainless Steel
STA.	Station
Std	Standard
STD	Stud
STL	Steel
STR	Straight
STRL	Structural
STRUC	Structural
SUB	Substitute
SUPP	Supplement
SUR	Surface
SUSP	Suspended
SV	Safety Valve
S/W, SW	Sidewalk
SWG	Standard Wire Gauge
SY	Square Yard
SYM	Symbol
SYM	Symmetrical
S2E	Surfaced on Two Edges
S2S	Surfaced on Two Sides
S4S	Surfaced on Four Sides

T	Thick
T	Threaded
T	Thermostat
TAN	Tangent
TB	Top of Beam
T&B	Top & Bottom
TBM	Temporary Bench Mark
TC	Top of Concrete
TC	Top of Curb
TEL	Telephone
TEMP	Temporary
TEMPL	Template
TERM	Terminal
TF	Top of Footing
TG	Top of Grate
T&G	Tongue & Grove
TH	Threshold
THK	Thickness
TM	Top of Masonry
TOF	Top of Footing
TP	Top of Pier
TRANS	Transformer
TS	Tensile Strength
TS	Time Switch
TS	Top of Slab
TV	Television
TW	Top of Wall
TYP	Typical
U	Up
UBC	Uniform Building Code
UG	Underground
UH	Unit Heater
UL	Underwriters Laboratory (electrical fixture specs)
UNEX	Unexcavated
UNF	Unfinished
UON	Unless Otherwise Noted
UR	Utility Room
USS	United States Standard
V	Valve
V	Vent
V	Volts
VAC	Vacuum Line
VD	Vent Duct
VAP PRF	Vapor proof
VC	Vitrified Clay

VEN	Veneer
VENT	Ventilate
VERT	Vertical
VEST	Vestibule
VI	Central Valve (as numbered)
VLF	Vertical Linear Feet
VOL	Volume
VP	Vitreous Pipe
VS	Vent Stack
VT	Vertical Tangent
VT&G	Vertical Tongue & Groove
VTR	Vent Through Roof
W	West
W	Width
W	Water
W/	With
WC	Water Closet
WD	Wood
WDW	Window
WF	Water Fountain
WGL	Wire Glass (reinforced)
WH	Water Heater
WHR	Watt Meter Hour
WI	Wrought Iron
WM	Water Meter
WP	Water Proof
WP	Weather Proof
WR BD	Weather Resistant Board
WS	Weather Stripping
WS	Water Service
WSCT	Wainscot
WT	Weight
WV	Water Valve
WWF	Welded Wire Fabric
X	By (multiplied by, as in 2 x 4)
X SECT	Cross Section (blueprint view)
XT	Temporary Marker
YD	Yard
YD^3	Cubic Yard
YP	Yellow Pine
Z	Zinc

Appendix

B

Industry Associations

The following associations are extremely important resource centers for the professional project scheduler. Most have information brochures, literature, and advice that they will forward at no cost.

The American Institute of Architects
1735 New York Ave.
Washington, DC 20006
202/626-7300

American National Standards Institute
11 W. 42nd Street
New York, NY 10036
212/642-4900

American Road & Transportation Assn.
501 School St., SW, 8th Floor
Washington, DC 20024
202/488-2722

American Society of Civil Engineers
345 E. 47th St.
New York, NY 10017
212/705-7496

American Subcontractors Assn.
1004 Duke St.
Alexandria, VA 22314
703/684-3450

Associated General Contractors
1957 E St., NW
Washington, DC 20006
202/393-2040

American Institute of Constructors
9887 N. Gandy, Ste. 104
St. Petersburg, FL 33702
813/578-0317

American Public Works Association
1313 E. 60th Street
Chicago, IL 60637
312/667-2200

American Society Testing Materials
(ASTM)
1916 Race St.
Philadelphia, PA
215/299-5400

American Society Home Inspectors
1735 N. Lynn St., Ste. 950
Arlington, VA 22209
800/296-2744

Associated Builders & Constructors
729 15th St., NW
Washington, DC 20005
202/637-8800

Building Officials Administrators
4051 W. Flossmoor Rd.
Country Club Hills, IL 60478
708/799-2300

Building Research Board
2101 Constitution Ave., NW
Washington, DC 20418
202/334-3376

Construction Management Institute
P.O. Box 1001
Soquel, CA 95073
408/462-0147

Construction Specifications Institute
601 Madison St.
Alexandria, VA 22314
703/684-0300

Society of Professional Engineers
1420 King St.
Alexandria, VA 22314
703/684-2800

Scaffolding Institute, Inc.
1300 Sumner Ave.
Cleveland, OH 44115
216/241-7333

Assn. of Geoscience Engineering
8811 Colesville Rd.
Silver Spring, MD 20907
301/565-2733

American Concrete Institute
22400 W. Seven Mile Rd.
Detroit, MI 48219
313/532-2600

Portland Cement Association
5420 Old Orchard Rd.
Stokie, IL 60077
708/966-6200

Wire Reinforcement Institute
1101 Connecticut Ave., NW
Washington, DC 20036
202/429-4303

Concrete Reinforcing Institute
933 N Plum Grove Rd.
Schaumberg, IL 60173
708/517-1200

National Precast Concrete Institute
825 E. 64th St.
Indianapolis, IN 46220
317/253-0486

Prestressed Concrete Institute
175 W. Jackson Blvd., Ste. 1859
Chicago, IL 60604
312/786-0300

Tilt-Up Concrete Association
2431 W. Cummings Wood Lane
Hendersonville, NC 28739
704/891-9578

International Masonry Institute
823 15th St., NW
Washington, DC 20005
202/783-3908

Masonry Institute of America
2550 Beverly Blvd.
Los Angeles, CA 90057
213/388-0472

Brick Institute of America
11490 Commerce Park Dr.
Reston, VA 22091
703/620-0010

Building Stone Institute
P.O. Box 5047
White Plains, NY 10602
914/232-5725

Cast Stone Institute
Greentree Pavilons, Ste. 408
Marlton, NJ 08053
609/858-0271

Limestone Institute of America
Stone City Bank Bldg., Ste. 400
Bedford, IN 47421
812/275-4426

Marble Center Assn.
499 Park Ave.
New York, NY 10022
212/980-1500

Marble Institute of America
33505 State St.
Farmington, MI 48335
313/476-5558

Aluminum Association
900 19th St., NW, Ste. 300
Washington, DC 20006
202/862-5100

American Institute of Steel Construction
1 E. Wacker Dr., Ste. 3100
Chicago, IL 60601
312/670-2400

American Iron & Steel Institute
1133 15th St., NW
Washington, DC 20005
202/452-7100

Copper Development Association
Greenwich Office Park 2
Greenwich, CT 06836
203/625-8210

Lath/Steel Framing Association
600 S. Federal St., Ste. 400
Chicago, IL 60606
312/922-6222

National Metal Manufacturers
600 S. Federal St., Ste. 400
Chicago, IL 60606
312/922-6222

Steel Structures Council
4400 5th Ave.
Pittsburgh, PA 15213
412/268-3327

Steel Joist Institute
1205 48th Ave., N, Ste. A
Myrtle Beach, SC 29577
803/449-0487

Steel Deck Institute
P.O. Box 9506
Canton, OH 44711
216/493-7866

Southern Forest Products Assn.
P.O. Box 52468
New Orleans, LA 70152
504/443-4464

Western Wood Products Assn.
522 SW 5th Ave.
Portland, OR 97204
503/224-3930

American Institute of Timber Construction
11818 SE Mill Plain Blvd.
Vancouver, WA 98684
206/254-9132

American Plywood Association
P.O. Box 11700
Tacoma, WA 98411
206/565-6600

Architectural Woodwork Institute
P.O. Box 1170
Centerville, VA 22020
703/222-1100

National Particleboard Association
2310 S. Walter Reed Dr.
Gaithersburg, MD 20879
301/670-0604

American Wood Preservers Assn.
P.O. Box 5283
Springfield, VA 22150
703/339-6660

Laminate Products Association
600 S. Federal St., Ste. 400
Chicago, IL 60606
312/922-6222

Sealant & Waterproofing Assn.
3101 Broadway, Ste. 585
Kansas, MO 64111
816/561-8230

Perlite Institute
88 New Dorp Plaza
Staten Island, NY 10306
718/351-5723

Exterior Insulation Association
2759 State Rd. 580, Ste. 112
Clearwater, FL 34621
813/231-6477

Cedar Shake & Shingle Assn.
515 116th Ave., NE, Ste. 275
Bellevue, WA 98004
206/453-1323

American Architectural Assn.
1540 E. Dundee Rd., Ste. 310
Palatine, IL 60067
708/202-1350

Vinyl Window & Door Institute
355 Lexington Ave.
New York, NY 10017
212/351-5400

Door & Hardware Institute
7711 Old Springhouse Rd.
McLean, VA 22102
703/556-3990

Cultured Marble Institute
435 N. Michigan Ave.
Chicago, IL 60611
312/644-0828

Institute of Roofing
4242 Kirchoff Rd.
Rolling Meadows, IL 60008
708/991-9292

Insulation Manufacturers Assn.
1420 King St., Ste 410
Alexandria, VA 22314
703/684-0084

Polyisocyanurate Insulation Assn.
1001 Pennsylvania Ave., NW
Washington, DC 20004
202/624-2709

Asphalt Roofing Association
6288 Montrose Rd.
Rockville, MD 20852
301/231-9050

Single Ply Roofing Institute
104 Wilmot Rd., Ste. 201
Deerfield, IL 60015
708/940-8800

National Wood Assn.
1400 E. Touhy Ave.
Des Plaines, IL 60018
708/299-5200

Steel Window Institute
1621 Euclid St.
Cleveland, OH 44115
216/241-7333

Steel Door Institute
30200 Detroit Rd.
Cleveland, OH 44145
216/899-0010

National Glass Association
8200 Greensboro Dr.
McLean, VA 22102

International Institute of Plaster
820 Transfer Rd.
St. Paul, MN 55111

Gypsum Association
810 1st St., Ste. 510
Washington, DC 20002
202/289-5440

Ceramic Tile Institute
700 North Virgil Ave.
Los Angeles, CA 90029
213/660-1911

Tile Council of America
P.O. Box 326
Princeton, NJ 08542
609/921-7050

National Mosaic Tile Assn.
3166 Des Plaines Ave.
Des Plaines, IL 60018
708/635-7744

Acoustical Society of America
500 Sunnyside Blvd.
Woodbury, NY 11797
516/349-7800

Interior Systems Association
104 Wilmot Rd.
Deerfield, IL 60015
708/940-8800

National Wood Flooring Assn.
11046 Manchester Rd.
St. Louis, MO 63122
800/422-4556

Resilient Floor Institute
966 Hungerford Dr.
Rockville, MD 20850
301/340-8580

National Coatings Association
1500 Rhode Island Ave.
Washington, DC 20005
202/462-6272

Wallcovering Manufacturers Assn.
355 Lexington Ave.
New York, NY 10017
212/661-4261

Plumbing Manufacturers Institute
800 Roosevelt Rd.
Glen Ellyn, IL 60137
708/858-9172

Air Diffusion Council
111 E. Wacker Dr., Ste. 200
Chicago, IL 60601
312/616-0800

American Society of Heating, Refrigerating and Air Conditioning Engineers (ASHRAE)
1791 Tullie Circle, NE
Atlanta, GA 30329
404/636-8400

Cooling Tower Institute
P.O. Box 73373
Houston, TX 77272
713/583-4087

Air Movement Control Assn.
30 W. University Dr.
Arlington Heights, IL 60004
703/394-0150

Institute of Heating & Air Conditioning (IHACI)
606 Larchmont Blvd., Ste. 4A
Los Angeles, CA 90004
213/467-1158

Edison Electric Institute
701 Pennsylvania Ave., NW
Washington, DC 20004
202/508-5000

Lighting Research Center
Polytechnic Institute, Bldg. No. 115
Troy, NY 12180
518/276-8716

National Electrical Contractors Assn
7315 Wisconsin Ave.
Bethesda, MD 20814
301/657-3110

Appendix

C

EJCDC Table of Contract Standards

The following appendix contains the table of standards set for general conditions of the construction contract that is prepared by the Engineers Joint Contract Documents Committee, recognized nationwide as the authoritative industry standard for construction contracts. Your project schedule will operate within the parameters of these standards. Use this table to locate the contracting and scheduling precedence of the item you seek, when and if there's a contract dispute delaying your schedule. Even if the project owner doesn't know this, you should. This table is of the current 1995 edition of the Standard General Conditions Of The Construction Contract, prepared by the EJCDC.

Article Number	Division	Title	Page Number
1.1	Contract Definitions	Addenda	13
1.2	Contract Definitions	Agreement	13
1.3	Contract Definitions	Payment Application	13
1.4	Contract Definitions	Asbestos	13
1.5	Contract Definitions	Bid	13
1.6	Contract Definitions	Bid Documents	13
1.7	Contract Definitions	Bid Requirements	13
1.8	Contract Definitions	Bonds	13
1.9	Contract Definitions	Change Order	13
1.10	Contract Definitions	Contract Documents	13
1.11	Contract Definitions	Contract Price	13
1.12	Contract Definitions	Contract Times	13
1.13	Contract Definitions	Contractor	13
1.14	Contract Definitions	Defective	13
1.15	Contract Definitions	Drawings	13
1.16	Contract Definitions	Effective Agreement Date	13
1.17	Contract Definitions	Engineer	13
1.18	Contract Definitions	Engineer's Consultant	13
1.19	Contract Definitions	Field Order	13
1.20	Contract Definitions	General Requirements	14
1.21	Contract Definitions	Hazardous Waste	14
1.22	Contract Definitions	Laws and Regulations	14
1.23	Contract Definitions	Liens	14
1.24	Contract Definitions	Milestones	14

Article Number	Division	Title	Page Number
1.25	Contract Definitions	Notice of Award	14
1.26	Contract Definitions	Notice to Proceed	14
1.27	Contract Definitions	Owner	14
1.28	Contract Definitions	Partial Utilization	14
1.29	Contract Definitions	PCBs	14
1.30	Contract Definitions	Petroleum	14
1.31	Contract Definitions	Project	14
1.32	Contract Definitions	Radioactive Material	14
1.33	Contract Definitions	Resident Representative	14
1.34	Contract Definitions	Samples	14
1.35	Contract Definitions	Shop Drawings	14
1.36	Contract Definitions	Specifications	14
1.37	Contract Definitions	Subcontractor	14
1.38	Contract Definitions	Substantial Completion	14
1.39	Contract Definitions	Supplementary Conditions	14
1.40	Contract Definitions	Supplier	14
1.41	Contract Definitions	Underground Facilities	14
1.42	Contract Definitions	Unit Price Work	14
1.43	Contract Definitions	Work	15
1.44	Contract Definitions	Work Change Directive	15
1.45	Contract Definitions	Written Amendment	15
2.0	Preliminary Matters	Delivery of Bonds	15
2.1	Preliminary Matters	Copies of Documents	15
2.2	Preliminary Matters	Contract Commencement	15
2.4	Preliminary Matters	Starting Work	15
2.5	Preliminary Matters	Preliminary Schedule	15
2.7	Preliminary Matters	Preconstruction Conference	15
2.9	Preliminary Matters	Accepted Schedules	16
3.0	Contract Documents	Standards	16
3.1 - 3.2	Contract Documents	Intent	16
3.3	Contract Documents	Specifications	16
3.4	Contract Documents	Terms & Adjectives	17
3.5	Contract Documents	Amending	17
3.6	Contract Documents	Supplementing Documents	17
3.7	Contract Documents	Reuse of Documents	17
4.0	Physical Conditions	Reference Points	17
4.1	Physical Conditions	Land Availability	17
4.2	Physical Conditions	Subsurface Conditions	17
4.2.1	Physical Conditions	Reports & Drawings	17
4.2.2	Physical Conditions	Reliance by Contractor	18
4.2.3	Physical Conditions	Differing Conditions Notice	18

Article Number	Division	Title	Page Number
4.2.4	Physical Conditions	Engineer's Review	18
4.2.5	Physical Conditions	Possible Contract Change	18
4.2.6	Physical Conditions	Possible Price Change	18
4.2.8	Physical Conditions	Possible Time Adjustments	18
4.3	Physical Conditions	Underground Facilities	18
4.3.1	Physical Conditions	Shown or Indicated	18
4.3.2	Physical Conditions	Not Shown or Indicated	19
4.4	Physical Conditions	Reference Points	19
4.5	Physical Conditions	Hazardous Materials	19
5.0	Bonds & Insurance	Bonds	20
5.1	Bonds & Insurance	Payment Bonds	20
5.2	Bonds & Insurance	Performance Bonds	20
5.3	Bonds & Insurance	Sureties	20
5.4	Bonds & Insurance	Contractor's Insurance	20
5.5	Bonds & Insurance	Owner's Insurance	21
5.6	Bonds & Insurance	Property Insurance	21
5.7	Bonds & Insurance	Machinery & Equipment	21
5.8	Bonds & Insurance	Notice of Cancellation	21
5.9	Bonds & Insurance	Contractor's Responsibility	22
5.10	Bonds & Insurance	Special Insurance	22
5.11	Bonds & Insurance	Waiver of Rights	22
5.13	Bonds & Insurance	Application & Receipt	22
5.14	Bonds & Insurance	Acceptance of Bonds	22
5.15	Bonds & Insurance	Partial Utilization	23
6.0	GC Responsibility	Supervision	23
6.1	GC Responsibility	Superintendence	23
6.4	GC Responsibility	Labor & Materials	23
6.5	GC Responsibility	Equipment	23
6.6	GC Responsibility	Progress Schedule	23
6.7	GC Responsibility	Item Substitutes	23
6.8	GC Responsibility	Engineer's Evaluation	23
6.11	GC Responsibility	Subcontractors	24
6.12	GC Responsibility	Fees	24
6.13	GC Responsibility	Permits	25
6.14	GC Responsibility	Laws & Regulations	25
6.15	GC Responsibility	Taxes	25
6.16	GC Responsibility	Use of Premises	26
6.17	GC Responsibility	Site Cleanliness	26
6.18	GC Responsibility	Safe Structural Loading	26
6.19	GC Responsibility	Record Documents	26
6.20	GC Responsibility	Safety & Protection	26
6.21	GC Responsibility	Safety Representative	26

Article Number	Division	Title	Page Number
6.22	GC Responsibility	Hazard Programs	27
6.23	GC Responsibility	Shop Drawings	27
6.24	GC Responsibility	Emergencies	27
6.25	GC Responsibility	Submittal Procedures	27
6.26	GC Responsibility	Engineer's Review	27
6.27	GC Responsibility	Variation from Contract	27
6.28	GC Responsibility	Related Work	27
6.29	GC Responsibility	Continuing Work	28
6.30	GC Responsibility	Warranty of Work	28
6.33	GC Responsibility	Indemnification	28
6.34	GC Responsibility	Obligations	28
7.0	Other Work	Related Work	29
7.2	Other Work	Site Work	29
7.4	Other Work	Coordination	29
8.0	Owr. Responsibility	Owner's Responsibility	29
8.1	Owr. Responsibility	Communication	29
8.2	Owr. Responsibility	Replacement of Engineer	29
8.3	Owr. Responsibility	Furnish Data	29
8.4	Owr. Responsibility	Easements	29
8.5	Owr. Responsibility	Insurance	29
8.6	Owr. Responsibility	Change Orders	29
8.7	Owr. Responsibility	Inspections & Tests	29
8.8	Owr. Responsibility	Stop Work	29
8.9	Owr. Responsibility	Limitations of Liability	30
8.10	Owr. Responsibility	Hazardous Waste	30
8.11	Owr. Responsibility	Financial Evidence	30
9.0	Engineer Status	During Construction	30
9.1	Engineer Status	Owner's Agent	30
9.2	Engineer Status	Site Visits	30
9.3	Engineer Status	Project Representative	30
9.4	Engineer Status	Interpretations	30
9.5	Engineer Status	Authorized Variations	30
9.6	Engineer Status	Rejecting Defective Work	30
9.7 - 9.9	Engineer Status	Shop Drawings	31
9.10	Engineer Status	Unit Prices Determination	31
9.11 - 9.12	Engineer Status	Dispute Decisions	31
9.13	Engineer Status	Limitations of Authority	31
10.0	Work Changed	During Construction	32
10.1	Work Changed	Owner Ordered	32
10.2	Work Changed	Adjustment Claim	32

Article Number	Division	Title	Page Number
10.3	Work Changed	Work Not Required	32
10.4	Work Changed	Change Orders	32
10.5	Work Changed	Surety Notification	32
11.0	Contract Changes	Price	32
11.2	Contract Changes	Claim for Adjustment	32
11.4	Contract Changes	Cost of Work	33
11.5	Contract Changes	Exclusions	34
11.6	Contract Changes	Contractor's Fee	34
11.7	Contract Changes	Cost Records	34
11.8	Contract Changes	Cash Allowances	35
11.9	Contract Changes	Unit Price Work	35
12.0	Changes in Times	Contract Time	35
12.1	Changes in Times	Claim for Adjustment	35
12.2	Changes in Times	Time of the Essence	35
12.3	Changes in Times	Contractor Delay	35
12.4	Changes in Times	Owner Delay	35
13.0	Tests & Inspections	Acceptance	36
13.1	Tests & Inspections	Notice of Defects	36
13.2	Tests & Inspections	Access to the Work	36
13.3	Tests & Inspections	Contractor Cooperation	36
13.4	Tests & Inspections	Owner Responsibility	36
13.5	Tests & Inspections	Contractor Responsibility	36
13.7	Tests & Inspections	Covering Work	36
13.9	Tests & Inspections	Uncovering Work	36
13.10	Tests & Inspections	Owner Stop Work	36
13.11	Tests & Inspections	Correction of Work	37
13.12	Tests & Inspections	Correction Period	37
13.13	Tests & Inspections	Acceptance of Work	37
13.14	Tests & Inspections	Owner Correction	37
14.0	Payments	To Contractor	37
14.1	Payments	Schedule of Values	37
14.2	Payments	Progress Payments	38
14.3	Payments	Contractor's Warranty	38
14.4 - 14.7	Payments	Review of Payments	39
14.8	Payments	Substantial Completion	39
14.10	Payments	Partial Utilization	39
14.11	Payments	Final Inspection	39
14.12	Payments	Final Application	40
14.14	Payments	Final Payment	40

Article Number	Division	Title	Page Number
14.15	Payments	Payment Acceptance	40
14.16	Payments	Waiver of Claim	40
15.0	Work Suspension	Termination	40
15.1	Work Suspension	Owner Suspension	40
15.2 - 15.4	Work Suspension	Owner Termination	40
15.5	Work Suspension	Contractor Termination	41
16.0 - 16.9	Dispute Resolution	Precedence	41
17.0	Miscellaneous	Extras	42
17.1	Miscellaneous	Giving Notice	42
17.3	Miscellaneous	Time Computation	42
17.4	Miscellaneous	Notice of Claim	42
17.5	Miscellaneous	Cumulative Remedies	42
17.6	Miscellaneous	Professional Fees	42
17.18	Miscellaneous	Court Costs	42
18.0	Acceptance	Bonds & Insurance	43
18.1	Acceptance	Defective Work	43
18.2	Acceptance	Final Payment	43
18.3	Acceptance	Insurance	43
18.4	Acceptance	Other Work	43
18.5	Acceptance	Substitutes	43
18.6	Acceptance	Work By Owner	44
19.0	Access	Lands	44
19.1	Access	Site	44
19.3	Access	Work	44
20.0	Acts/Omissions	Contractor	45
20.1	Acts/Omissions	Engineer	45
20.2	Acts/Omissions	Architect	45
20.3	Acts/Omissions	Owner	45
20.4	Acts/Omissions	Inspector	45
21.0	Addenda	Definition	46
21.1	Addenda	Specifications	46
21.2	Addenda	Insurance	46
21.3	Addenda	Adjustments	46
21.5	Addenda	Prices	46
21.6	Addenda	Contract Times	46
21.7	Addenda	Adjustments	46
21.8	Addenda	Progress	47

Article Number	Division	Title	Page Number
21.9	Addenda	Agreement	47
22.0	Amendment	Written	47
22.1	Amendment	General	47
22.2	Amendment	Claims	47
22.3	Amendment	Contractor	47
22.4	Amendment	Owner	47
22.5	Amendment	Architect	48
22.6	Amendment	Engineer	48
22.8	Amendment	Definition	48
22.9	Amendment	Prices	48
23.0	Authorization	Owner	48
23.1	Authorization	Architect	48
23.2	Authorization	Engineer	49
23.3	Authorization	Contractor	49
23.4	Authorization	Prior	49
23.5	Authorization	During Construction	49
23.6	Authorization	Change Order	49
23.7	Authorization	Claims	49
24.0	Contract Specs	Documents	50
24.1	Contract Specs	Bidding	50
24.2	Contract Specs	Award	50
24.3	Contract Specs	Recordation	50
24.4	Contract Specs	Bonds & Insurance	50
24.5	Contract Specs	Amendment	50
24.6	Contract Specs	Arbitration	50
24.7	Contract Specs	Changes	51
24.8	Contract Specs	Change Order	51
24.9	Contract Specs	Claims	51
25.0	Certificates	Bonds	51
25.1	Certificates	Insurance	51
25.2	Certificates	Cancellation	51
25.3	Certificates	Cash Allowance	52
25.5	Certificates	Substantial	52
25.6	Certificates	Inspections	52
25.7	Certificates	Final	52
25.8	Certificates	Completion	52
25.9	Certificates	Cancellation	52

Article Number	Division	Title	Page Number
26.0	Contractor	Defined	53
26.1	Contractor	Fee	53
26.2	Contractor	Cost of Work	53
26.3	Contractor	General	53
26.4	Contractor	Exclusions	53
26.5	Contractor	Cost Records	53
26.6	Contractor	Lump Sum	53
26.7	Contractor	Unit Price	53
26.8	Contractor	Surety Notify	54
26.9	Contractor	Scope of	54
27.0	Owner	Defined	54
27.1	Owner	General	54
27.2	Owner	Scope of	54
27.3	Owner	Insurance	54
27.4	Owner	Coordination	55
27.5	Owner	Disclosures	55
27.6	Owner	Value of Work	55
27.7	Owner	Schedule	55
27.8	Owner	Testing	55
27.9	Owner	Inspections	55
28.0	Architect	Defined	55
28.1	Architect	General	56
28.2	Architect	Scope of	56
28.3	Architect	Coordination	56
28.5	Engineer	Defined	56
28.6	Engineer	General	56
28.7	Engineer	Scope of	56
28.8	Engineer	Coordination	56
29.0	Change Orders	Defined	57
29.1	Change Orders	Acceptance	57
29.2	Change Orders	Amendments	57
29.3	Change Orders	Cash Allowance	57
29.4	Change Orders	Price	57
29.5	Change Orders	Contract Times	57
29.6	Change Orders	Work Changes	57
29.7	Change Orders	Schedule	57
30.0	Work Changes	Responsibilities	58
30.1	Work Changes	Execution of	58
30.2	Work Changes	Indemnification	58
30.3	Work Changes	Insurance	58

Article Number	Division	Title	Page Number
30.4	Work Changes	Bonds	58
30.5	Work Changes	Terminate	58
30.6	Work Changes	Recordation	58
30.7	Work Changes	Substitutes	58
30.8	Work Changes	Unit Price	58
30.9	Work Changes	Lump Sum	58
31.0	Claims	Contractor	59
31.1	Claims	Owner	59
31.2	Claims	Architect	59
31.3	Claims	Engineer	59
31.4	Claims	Price Changes	59
31.5	Claims	Contract Times	60
31.6	Claims	Contractor's Fee	60
31.7	Claims	Liability	60
31.8	Claims	Cost of Work	60
31.9	Claims	Adjustments	60
32.0	Disputes	Resolution	61
32.1	Disputes	Agreement	61
32.2	Disputes	Precedence	61
32.3	Disputes	Arbitration	61
32.4	Disputes	Arbitrator	61
33.0	Conditions	Amending	62
33.1	Conditions	Bonds	62
33.2	Conditions	Cash Allowances	62
33.3	Conditions	Price Changes	62
33.4	Conditions	Time Changes	62
33.5	Conditions	Work Changes	62
33.6	Conditions	Clarifications	62
33.7	Conditions	Interpretation	63
33.8	Conditions	Definition	63
33.9	Conditions	Agreement	63
34.0	Contract Price	Bid	63
34.1	Contract Price	Amendments	63
34.2	Contract Price	Adjustments	63
34.3	Contract Price	Change of	63
34.4	Contract Price	Definition of	63
34.5	Contract Price	Acceptance	64
34.6	Contract Price	Cash Allowances	64
34.7	Contract Price	Payment Schedule	64

Article Number	Division	Title	Page Number
34.8	Contract Price	Progress Payments	64
34.9	Contract Price	Final Payment	64
35.0	Subcontractor	Defined	65
35.1	Subcontractor	Responsibility	65
35.2	Subcontractor	Schedule	65
35.3	Subcontractor	Communications	65
35.4	Subcontractor	Cooperation	65
35.5	Subcontractor	Continue Work	65
35.6	Subcontractor	Stop Work	65
35.7	Subcontractor	Shop Drawings	65
35.8	Subcontractor	Progress Schedule	66
35.9	Subcontractor	Progress Payment	66
36.0	Work Items	Defined	66
36.1	Work Items	Acceptance	66
36.2	Work Items	Standards	66
36.3	Work Items	Rejection	66
36.4	Work Items	Reinstallation	66
36.5	Work Items	Adjustments	67
36.6	Work Items	Continue Work	67
36.7	Work Items	Procurement	67
36.8	Work Items	Safety	67
36.9	Work Items	Storage	67
37.0	Safety	Defined	67
37.1	Safety	Representative	68
37.2	Safety	Program	68
37.3	Safety	Inspections	68
37.5	Safety	Reports	68
37.6	Safety	Recordation	68
37.7	Safety	Meetings	68
37.8	Safety	Site Cleanliness	68
37.9	Safety	Special Conditions	68
38.0	Cost	Defined	69
38.1	Cost	of Work	69
38.2	Cost	of Correction	69
38.3	Cost	Bonds & Insurance	69
38.4	Cost	Cash Discounts	69
38.5	Cost	Expenses	69
38.6	Cost	General	69
38.7	Cost	Exclusions	70
38.8	Cost	Office Overhead	70

Article Number	Division	Title	Page Number
38.8.2	Cost	Home Office	70
38.9	Cost	Employees	70
38.9.1	Cost	Labor	70
38.9.2	Cost	Labor Burden	70
38.9.3	Cost	Insurance	70
38.9.5	Cost	Losses	70
38.9.6	Cost	Materials	70
38.9.7	Cost	Equipment	71
38.9.8	Cost	Minor Expenses	71
38.9.9	Cost	Taxes	71
39.0	Covering Work	Contractor	71
39.1	Covering Work	Precedence	71
39.2	Covering Work	Schedule	71
39.3	Covering Work	Liens	71
39.4	Covering Work	Limitations	71
39.5	Covering Work	Definitions	71
39.6	Covering Work	Liability	71
39.7	Covering Work	Bonds & Insurance	72
39.8	Covering Work	Claims	72
39.9	Covering Work	Waiver of Claim	72
40.0	Laws	Bonds	72
40.1	Laws	Work Changes	72
40.2	Laws	Contract Documents	72
40.3	Laws	Work Rejection	72
40.4	Laws	Acceptance	72
40.5	Laws	Precedence	72
40.6	Laws	Reference to	72
40.7	Laws	Corrections	73
40.8	Laws	Indemnification	73
40.9	Laws	General	73
50.0	Equipment	Defined	73
50.1	Equipment	General	73
50.2	Equipment	Operations	73
50.3	Equipment	Safety & Protection	73
50.4	Equipment	Storage	74
50.5	Equipment	Maintenance	74
50.6	Equipment	Insurance	74
50.7	Equipment	Leased	74
50.8	Equipment	Mobilization	74
50.9	Equipment	Demobilization	74

Article Number	Division	Title	Page Number
60.0	Project	Defined	75
60.1	Project	Vested Parties	75
60.2	Project	Representative	75
60.3	Project	General	75
60.4	Project	Property	75
60.5	Project	Land	75
60.6	Project	Access	75
60.7	Project	Commencement	76
60.8	Project	Work in Progress	76
60.9	Project	Completion	76
60.9.1	Project	Closeout	76
61.1	Schedule	Defined	77
61.2	Schedule	General	77
61.3	Schedule	CPM	77
61.4	Schedule	Adherence to	77
61.5	Schedule	Progress	77
61.6	Schedule	Shop Drawings	77
61.7	Schedule	Adjusted	77
61.8	Schedule	As-Built	78
61.9	Schedule	Scope of Changes	78
61.9.3	Schedule	Work Stoppage	78
62.0	Supplier	Defined	79
62.1	Supplier	General	79
62.2	Supplier	Principals	79
62.3	Supplier	Substitutes	79
62.4	Supplier	Liens	79
62.5	Supplier	Waiver of Rights	79
62.6	Supplier	Shop Drawings	80
62.7	Supplier	Schedule	80
62.8	Supplier	Delivery	80
62.9	Supplier	Site Access	80
63.0	Submittals	Defined	81
63.1	Submittals	General	81
63.2	Submittals	Specifications	81
63.3	Submittals	Payment Application	81
63.4	Submittals	Maintenance	81
63.5	Submittals	Recordation	81
63.6	Submittals	Operations	81
63.7	Submittals	Procedures	82
63.8	Submittals	Review	82
63.9	Submittals	Authorization	82

Article Number	Division	Title	Page Number
64.0	Warranties	Defined	82
64.1	Warranties	of Work	82
64.2	Warranties	of Items	82
64.3	Warranties	of Documents	82
64.5	Warranties	of Record	82
64.6	Warranties	Waiver of	82
64.7	Warranties	Utilization	83
64.8	Warranties	Expressed	83
64.9	Warranties	Implied	83
70.0	Work	Defined	84
70.1	Work	General	84
70.2	Work	Access to	84
70.3	Work	by Others	84
70.4	Work	Continuing	84
70.5	Work	Delay	85
70.6	Work	Contractor Stoppage	85
70.6.3	Work	Owner Stoppage	85
70.6.5	Work	Engineer Stoppage	85
70.6.7	Work	Architect Stoppage	85
70.6.8	Work	Inspector Stoppage	85
70.7	Work	Coordination of	86
70.8	Work	Cost of	86
70.8.2	Work	Neglect of	86
70.8.3	Work	Related	86
70.8.5	Work	Variation	86
70.8.6	Work	Commencement	86
70.8.7	Work	Inspections	87
70.8.8	Work	Progress Schedule	87
70.8.9	Work	Final	87
70.9	Work	Completion	87
71.0	Progress	Defined	88
71.1	Progress	General	88
71.2	Progress	Schedule	88
71.3	Progress	Projected	88
71.4	Progress	Variation	88
71.5	Progress	Delay	88
71.6	Progress	Contractor Termination	88
71.7	Progress	Owner Termination	89
71.8	Progress	Weather	89
71.9	Progress	Completion	89

Index

A

abbreviations, project/production documents, 311-324
acceleration of schedule, 292-295
accounting and bookkeeping, 22, 276-286, 299
  accrual method, 276-277
  assets, 277-278, 279
  balance sheets, 277
  billing retention, 284-285
  capitalization, 284
  cash method, 276-277
  costs, 280
  debt, 285-286
  equity, 285
  financial statements, 277
  financing sources, 285
  income statement, 279-281
  liabilities, 277-278, 279
  profits, 286
  ratios, 281-284
  record keeping, 276
  rentention, 284-285
  Small Business Administration (SBA) loans, 285-286
accrual method accounting, 276-277
activities, 30
activity arrows, 51, **52**, 65, 67
activity duration, 30
activity functions, 158
activity lists, 30
activity numbers, 30
activity-on-node diagramming, 87-88, **89**
addressing, relative addressing, in software, 173
adhesion, contracts of adhesion, 297
adjusted schedules, 166
alignment of values and labels, 169
arbitration, in disputes, 106
arrows and arrow diagramming method (ADM), 26, 27, 30, 34, 83, **84**
  activity arrows, 51, **52**, 65, 67
  dummy arrows, 53, **54**, **55**, **56**, 65, 87
  I-J numbers, 53, **54**, **55**, **56**, 57, 65, 67, 83, **84**
  parallel activities diagrammed, **55**, **56**
  precedence diagramming vs., 87
as-built schedules, 166
as-planned schedules, 166
assets, 277-278, 279
associations, industry-related associations, 325-330
audit trail generation, 25, 26, 47-48, 139-140, 306
  Audit Trail Tracking, 244, **245**
  Schedule of Anticipated Earnings, 219, **220-222**
Audit Trail Tracking, 244, **245**

B

balance sheets, 277
Bar Chart by Early Start, 229, **230-233**
bar charts, 9, 13-16, **15**, 27, **38**, 46, 79
  Bar Chart by Early Start, 229, **230-233**
  CPM vs. bar charts, 82
  dependency bar chart scheduling, 88
benchmarking, 13
bidding and contractual documents, 143-144
  bid award prior to commencement, 152-153
billing, 22
billing retention, 284-285
blueprints, shop drawings, 147-149
boilerplate contracts, 143
bonded stop notice, 123-124
bonds, 274-275
bookkeeping (*see* accounting and bookkeeping)
bottlenecks, 65
breach of contract, 298
budgeting (*see* cost estimates)
business management, 273-275
  bonds, 274-275
  contract bonds, 275
  corporations, 274
  partnerships, 273
  payment bonds, 275
  performance bonds, 274
  permits, 274
  sole ownership of business, 273
  taxes, 274

C

cancelling commands, 249
capital stock, 279
capitalization, 284
cash method, 276-277
cell alignment, 254
cell references or cell addresses, 172, 173
cessation, notice of cessation, 118
chain-of-command and responsibility, 10-11
Change Order Tracking, 242, **243**

Illustrations are indicated by boldface.

change orders, 109-110, 163
  Change Order Tracking, 242, **243**
  pricing change orders, 297
changed (unforeseen) conditions, 292
changes (*see* contingency planning)
changing or editing data, 180-182
charting and scheduling, 9
  bar charts, 9
  S-charts, 9
  velocity diagrams, 9
claims against subcontractors for delays, 99, 100-104, 289
collective bargaining, 128, 301
columns, database operations, 185, 253-254
comma placement in numbers, 171
command entry, menu commands, 177-178
completion, notice of completion, 118, 132
computer assisted design (CAD), 22
computer requirements, 21-23
conflicts in project schedule, 297-303
constraints, 32, 35, 37, 45, 46, 67, 77, 87
construction law, 125-132
Construction Managers (CM), 1
constructive changes, 291
consultative capacity of project schedulers, 6
contingency planning, 42, 46, 289-303
  acceleration of schedule, 292-295
  change order pricing, 297
  changed (unforeseen) conditions, 292
  changes to schedule, 290-291
  claims against subcontractors for delays, 289
  conflicts in project schedule, 297-303
  constructive changes, 291
  delays, 289
  overtime losses, 295-296
  productivity losses, 295-296
  suspension of project schedule, 296-297
contract bonds, 275
Contract/P.O. Summary, 246, **247**
contracts (*see also* legal considerations), 6, 7, 10-11, 22, 113-124, 143-149
  avoiding problems, 110-111
  bid award prior to commencement, 152-153
  bidding and contractual documents, 143-144
  boilerplate contracts, 143
  breach of contract, 298
  change orders, 109-110, 163, 242, **243**
  conditions of the contract, 144
  Contract/P.O. Summary, 246, **247**
  contracts of adhesion, 297
  daily inspection reports (DIR), 116
  delays, 113-115
  EJCDC contract standards, 115, 331-343
  legal notices, 118-123, 132
  liens, timetable for filing, 132
  long-lead items purchase orders, 149
  mechanic's liens, 120-122
  no-damage clauses, 102-103, 117
  notice of cessation, 118
  notice of completion, 118, 132
  notice of non-responsibility, 118
  notice to owner, 122-123, 132
  outside delays, 115-117
  preliminary notices, 119-120, 132
  project schedulers and the law, 113
  quality assurance reports (QAR), 116
  shop drawings, blueprints, etc., 147-149
  specifications and standards, 144, 146
  stop notices, 123-124, 132
contracts of adhesion, 297
corporations, 274
Correspondence Transmittals, 240, **241**
Cost by Activity Number, 215, **216-218**
cost control/cost tracking, 25, 26, 43, 47-48, 138, 152, 156, 158-159, 280, 306
  audit trail generation, 25, 26, 47-48, 139-140, 306
  Cost by Activity Number, 215, **216-218**
  cost-loaded CPM, 97-98, 163-164
  fast-tracking, 91-92
  Schedule of Anticipated Earnings, 219, **220-222**
cost estimates, 2, 22, 40, 45, 47-48, 59, 135-140
  cost-loaded CPM, 97-98, 163-164
  long-lead items, 137
  quantity surveys, 136-137
  takeoffs, 135-136
  trends analysis, 139-140
cost of goods sold, 280
cost of operations, 280
cost of using CPM, 27-28, 82
cost tracking (*see* cost control/cost tracking)
CPM schedule, 32
criminal records of employees, 127, 301
critical path management (CPM), 14, 25-50
  arrow diagramming method (ADM), 26, 27, 34
  audit trail generation, 25, 26
  bar charts, 27
  benefits of CPM vs. bar charting, 46
  constraints, 35, 37
  cost estimates, 40, 47-48
  cost of using CPM, 27-28, 82
  cost tracking reports, 25, 26
  cost-loaded CPM, 97-98, 163-164
  critical areas of control in CPM, 25
  critical vs. non-critical activities, 26, 37, 65, 159
  delays, 26
  detailing the CPM project schedule, 44-46
  developing the CPM schedule, 33-39
  disadvantages of CPM, 82
  evaluating the CPM schedule plan, 40
  events, 27
  fast-tracking, 95-96
  float, 26, 28
  garage construction example, 36, **36**
  human resources, 48

interlinking or interdependence of tasks, 26, 27
job logic sequencing, 26-27
job logic, 41-46
matrix networking, 49-50
milestones, 40
network diagram preparation, 28, 35
opportunities assessment, 33
past-performance data, 33
performance targets, 41
precedence diagramming method (PDM), 26, 35
preconstruction planning, 34
productivity assessment, 33
project planning, 41-46
residential construction example of CPM, 37, **38**
S-charts, 25
scheduling phase, 28-29
terminology of CPM, 30-32
time management, 33
timeline computations, 28, 29, 35, 37
velocity diagrams, 29
vulnerabilities assessment, 33
critical vs. non-critical activities, 26, 37, 65, 159
cross-tabulation tables, 188
current assets, 277
current liabilities, 278
current ratio, 282
customer base building, 268-273

D

daily field reports, 160-162, 227, **228**
daily inspection reports (DIR), 101, 116
data entry for software, 167-170
data tables, 182, 188
data types, database operations, 185
database operations, 182-255
Audit Trail Tracking, 244, **245**
Bar Chart by Early Start, 229, **230-233**
building a database, 184-186
cancelling commands, 249
cell alignment, 254
Change Order Tracking, 242, **243**
closing a sort, 249
columns, 185, 253-254
Contract/P.O. Summary, 246, **247**
Correspondence Transmittals, 240, **241**
Cost by Activity Number, 215, **216-218**
cross-tabulation tables, 188
Daily Field Reports, 227, **228**
data tables, 182, 188
data types, 185
deleting cells, rows, columns, 251-252
directory of FastPro sorts, 190
double-key sort, 187
exiting FastPro, Lotus, Quattro, 250-251
extra keys sort, 187
field names, 185-186
fields, 183-184
file management, maintaining schedule sorts, 258-260
input cells, 182
input values, 182
key fields, 184
key sorts, 186-187
maximum number of records in database operations, 184-185
merging cells, 252-253
Network Timeline, 234, **235**
opening a sort, 250
printing, 255
ranges, data table ranges, 188-189
records, 183-184
results area, 182-183
row height setting, 254
saving sorts, 248-249
saving sorts to a different disk drive, 260-261
saving sorts to different subdirectory, 260
Schedule of Anticipated Earnings, 219, **220-222**
Shop Drawings Log, 236, **237**
Sort by Activities, 190, **191-194**
Sort by Early Starts, 223, **224-226**
Sort by Events, 195, **196-199**
Sort by I-J Numbers, 200, **201-204**
Sort by Job Logic, 205, **206-209**
Sort by Total Float/Late Start, 210, **211-214**
sorting databases, 186
splitting cells, 252-253
spreadsheet data into databases, 185
Submittal Items Tracking, 238, **239**
variables, 182
what-if analysis, 189
deadlines, 59
debt, 285-286
deductions for expenses, taxes, 268
delays, 26, 113-115, 155, 289
claims against subcontractors for delays, 99, 100-104
outside delays, 115-117
deleting cells, rows, columns, 251-252
dependencies, event dependencies, 59
dependency bar chart scheduling (*see also* activity-on-node), 88
design/build systems, 92-95
directory of FastPro sorts, database operations, 190
disadvantages of CPM, 82
discrimination, 128, 302
disk drives, saving sorts to a different disk drive, 260-261
disputes and litigation (*see also* contracts; legal considerations), 99-111
dollar amounts, 171
double-key sort, 187
downsizing tends, 307-309
dummy arrows, 53, **54**, **55**, **56**, 65, 87
durations factoring fast-track, **94**, **95**

E

early finish dates, 31
early start date, 31
  Bar Chart by Early Start, 229, **230-233**
  Sort by Early Starts, 223, **224-226**
earnings, retained earnings, 279
editing data, 180-182
employees (*see* human resources)
Engineers Joint Contract Documents Committee (EJCDC), 115
  contract standards, 115, 331-343
equity, 279, 285
errors and extras, 4-5
estimating (*see* cost estimates)
evaluating the CPM schedule plan, 40
event dependencies, 59
event diagrams, 31
events, 27, 30
  Sort by Events, 195, **196-199**
evolution and development of project scheduling, 2-3
exiting FastPro, Lotus, Quattro, 250-251
expenses, 280
  accrued expenses payable, 279
  prepaid expenses, 278
exponential factoring, 175
extra keys sort, 187

F

factoring, exponential factoring, 175
FastPro software use (*see also* database operations), 167-261
  alignment of values and labels, 169
  cancelling commands, 249
  cell alignment, 254
  changing or editing data, 180-182
  column width setting, 253-254
  comma placement in numbers, 171
  command entry, menu commands, 177-178
  data entry, 167-170
  database (*see* database operations)
  deleting cells, rows, columns, 251-252
  dollar amounts, 171
  exiting FastPro, Lotus, Quattro, 250-251
  exponential factoring, 175
  file management, maintaining schedule sorts, 258-260
  formulas, 170, 172-177
  global commands, 168
  I-J numbers, 168
  installation, 167
  labels, 168
  merging cells, 252-253
  numeric data, 169, 170-171
  object linking and embedding (OLE), 180
  percentages, 171
  printing, 255
  procedures, 180
  range commands, 168
  ranges, 180
  row height setting, 254
  saving your work, 178-179, 181-182
    saving sorts to a different disk drive, 260-261
    saving sorts to different subdirectory, 260
    saving sorts, 248-249
  scientific notation numbers, 171
  speed/shortcut commands, 179
  splitting cells, 252-253
  string operators, 175, **176**, 177
  subdirectories, 260
  syntax errors in data entry, 169-170
  too long numbers, 171
  Undo command, 181-182
  values, 168
  Windows use of FastPro, 255-258
    closing windows, 258
    copying and moving data between windows, 256-257
    moving between windows, 256
    moving windows, 257
    opening a window, 256
  WYSIWYG format, 168
fast-tracking, 78, 91-98, 159, 306
  cost control using fast-tracking, 91-92
  cost-loaded CPM, 97-98, 163-164
  critical path management (CPM) and fast-tracking, 95-96
  design/build systems, 92-95
  durations factoring fast-track, **94**, **95**
  long-lead items, 96
  phased construction methods, 91-92
  phases of a fast-tracked schedule, 93
  preconstruction conferences, 96
field names, database operations, 185-186
field scheduling, 19, 306-307
fields, database operations, 183-184
file management, maintaining schedule sorts, 258-260
finances (*see* accounting and bookkeeping)
financial statements, 277
financing sources, 285
finish dates, 31, 44, 65
firing employees, 125, 126
fixed assets, 278
float, 26, 28, 31, 46, 67, 69-76, 99
  claims against subcontractors for delays, 99, 100-104
  computing float, 73, **74**, 75, **76**
  constraints, 71-72
  critical vs. non-critical activities, 69
  distributed float, 71
  example of float, 72-75, **74**, **76**

free float, 31, 46, 65, 70, 72, 75
high float, 71
line float, 70
low float, 71
negative float, 31, 45, 71
penalties, cash penalties for late completion, 99
Sort by Total Float/Late Start, 75, **76**, 210, **211-214**
takeoffs, 135-136
total float, 31, 70, 72, 75
flowlines, 79-80, **81**
formulas in databases, 170, 172-177
cell references or cell addresses, 172, 173
exponential factoring, 175
length of formula, 172
logic formulas, 173, **174**
operators, 172
relative addressing, 173
string operators, 175, **176**, 177
free float, 31, 46, 65, 70, 72, 75

G

garage construction example of CPM, 36, **36**
global commands, 168
graphics requirements, 21

H

home office deduction, 267
hours of work, 127, 301
human resources management, 299-303
collective bargaining, 128, 301
criminal records of employees, 127, 301
discrimination, 128, 302
firing employees, 125, 126
hours of work, 127, 301
illegal aliens as employees, 128, 302
labor codes, 125-129
labor use, union vs. non-union labor, 125, 127, 301
medical examinations, 126, 300
minors as employees, 128, 302
motivating employees, 43, 156
Occupational Safety and Health Administration (OSHA), 125, 129-131
overtime pay, 127, 301
payment of wages due, 126-127, 299, 300
pension funds, 127, 300
polygraph tests, 127, 300-301
public works, 131-132
safety of employees, 125
sick leave, 127, 300
Social Security Numbers, 129, 302
strikes and work stoppages, 126, 127, 128, 300, 301
traits for success, 263-265
unemployment insurance payments, 129, 302-303
vacation pay, 127, 300
withholding taxes, 127, 300

I

I-J number, 31
I-J numbers, 31, 53, **54**, **55**, **56**, 57, 60, 65, 67, 78, 85, **86**, 88, **89**, 168
Sort by I-J Numbers, 200, **201-204**
illegal aliens as employees, 128, 302
income statement, 279-281
income, 280-281
incorporation, 274
industry-related associations, 325-330
input cells, database operations, 182
input values, database operations, 182
inspections, daily inspection reports (DIR), 101, 116
installing software, 167
integrated CPM software, 8, 9-10, 22-23, 306
interlinking or interdependence of tasks, 10, 26, 27
inventory, 278

J

job logic, 7, 26-27, 32, 41-46, **52**, 59-60, 77, 151, 305
I-J numbers, 60
logic-based scheduling, 63-67
job logic, 7, 26-27, 32, 41-46, **52**, 59-60, 77, 151, 305
I-J numbers, 60
logic-based scheduling, 63-67, **64**, **66**
loops, logic loops, 60
program logic, 61, **62**
Sort by Job Logic, 205, **206-209**

K

key fields, database operations, 184
key sorts, 186-187

L

labels, 168
labor (*see* human resources)
labor codes, 125-129
labor costs, 280
labor unions, 127, 301
laptop computers, 286-287
late finish dates, 31
penalties, cash penalties for late completion, 99
late start dates, 31
Sort by Total Float/Late Start, 210, **211-214**
legal considerations (*see also* contracts; human resources management), xv-xvi, 5, 10-11
arbitration, 106
avoiding problems, 110-111
chain-of-command and responsibility, 10-11

legal considerations *continued*
- change orders, 109-110, 163
- claims against subcontractors for delays, 99, 100-104
- collective bargaining, 128, 301
- construction law, 125-132
- contracts, 10-11, 113-124, 143-149
- criminal records of employees, 127, 301
- daily inspection reports (DIR), 101, 116
- delay, 113-115
- discrimination, 128, 302
- disputes and litigation, 99-111
- firing employees, 125, 126
- hours of work, 127, 301
- illegal aliens as employees, 128, 302
- labor codes, 125-129
- labor unions, 125, 127, 301
- legal notices, 118-123
- lien rights, 10-11
- liens, timetable for filing, 132
- litigation, 106, 107-108
- mechanic's liens, 120-122
- mediation, 105-106
- medical examinations, 126, 300
- minors as employees, 128, 302
- multiple critical path, 108-109
- no-damage clauses, 117
- Occupational Safety and Health Administration (OSHA), 125, 129-131
- outside delays, 115-117
- overtime pay, 127, 301
- payment of wages due, 126-127, 299-300
- penalties, cash penalties for late completion, 99
- pension funds, 127, 300
- polygraph tests, 127, 300-301
- project schedulers and the law, 113
- public entities and no-damage clauses, 102-103
- public works, 131-132
- quality assurance reports (QAR), 101, 116
- safety of employees, 125
- sick leave, 127, 300
- Social Security Numbers, 129, 302
- strikes and work stoppages, 126, 127, 128, 300, 301
- unemployment insurance payments, 129, 302-303
- vacation pay, 127, 300
- vested parties, 10
- withholding taxes, 127, 300

legal notices, 118-123, 132
length of entries in software, 171
liabilities, financial, 277-279
licensing, 5, 274
lien rights, 10-11, 120-122, 132
line float, 70
litigation, 106, 107-108
local area networks (LAN), 22-23
logic diagram, 32
logic formulas, 173, **174**
logic loops, 60
logic-based scheduling, 63-67, **64**, **66**
logic-diagram chart scheduling, 13
long-lead items, 45, 96, 137, 306
- purchase orders, 149

long-term liabilities, 279
loops, logic loops, 60

M

macroactivities vs. microgrouping, 14, 16, 59
marketing your services, customer base building, 268-273
matrix networking, 49-50
mechanic's liens, 120-122
mediation, in disputes, 105-106
medical examinations, 126, 300
memory requirements for computer, 21-22
merging cells, 252-253
microgrouping vs. macroactivities, 14, 16
milestones, 13, 22, 31, 40, 42, 45, 57-59, 151, 155
minors as employees, 128, 302
monitoring aspects of construction, 7
multiple critical path, 78, 108-109
multiuser server functionality, networks, 23

N

negative float, 31, 45
net income to total assets ratio, 284
net profits, 281
net sales to net working capital ratio, 283
net sales to net worth ratio, 283-284
network diagram preparation, 28, 35, **39**, 51-67
- activity arrows, 51, **52**, 65, 67
- bottlenecks, 65
- constraints, 67
- critical vs. non-critical activities, 65
- deadlines, 59
- dummy arrows, 53, **54**, **55**, **56**, 65
- event dependencies, 59
- finish dates, 65
- float, 67
- free float, 65
- I-J numbers, 53, **54**, **55**, **56**, 57, 60, 65, 67
- job logic, **52**, 59-60
- logic loops, 60
- logic-based scheduling, 63-67, **64**, **66**
- milestones, 57-59
- parallel activities diagrammed, **55**, **56**, 63
- program logic, 61, **62**
- resource allocation, 59
- start dates, 65, 67
- subtasks, 63
- tasks, 63
- timescale computations, 65, 67

network (communications) requirements, 22-23
Network Timeline, 234, **235**
no-damage clauses, 102-103, 117
non-responsibility, notice of non-responsibility, 118
notice of cessation, 118
notice of completion, 118, 132
notice of non-responsibility, 118
notice to owner, 122-123, 132
numeric data, 169, 170-171

O

object linking and embedding (OLE), 180
Occupational Safety and Health Administration (OSHA), 125, 129-131
on site scheduling (*see* field scheduling), 19
opening a sort, 250
operators, 172
overtime losses, 3, 295-296
overtime pay, 3, 127, 301

P

parallel activities diagrammed, **55**, **56**, 63
partnerships, 273
payment bonds, 275
payroll, 279
pension funds, 127, 300
percentages, 171
performance bonds, 274
performance targets, 41
permits, 274
personnel (*see* human resources)
PERT chart scheduling, 13, 16-17, 32
phase scheduling, 77
phased construction (*see* fast-tracking)
philosophy of scheduling, 18
polygraph tests, 127, 300-301
portable computers, 286-287
precedence diagramming, 26, 32, 35, 83, 85, **86**, 87
preconstruction planning, 34, 151-156
  bid award prior to commencement, 152-153
  cost control, 152, 156
  delays, 155
  developing schedules, 156
  fast-tracking, 96
  job logic, 151
  milestones, 151, 155
  motivating employees, 156
  priortizing, 155
  problems, 155
  profitability, 155
  regimentation of progress schedule, 151-152, 153
  S-charts, 151
  schedule planning, 155
  shop drawings log, 154
  timescales, 155
  transmittals, 154-155
preliminary notices, 119-120, 132
prepaid expenses, 278
printing, 255
prioritizing, 42, 155
procedures, 180
Production Coordinator (PC), 1
productivity, 7, 33, 295-296
profitability, 4-5, 6, 9, 42, 155, 281, 286
  Schedule of Anticipated Earnings, 219, **220-222**
Program Evaluation Review Technique (*see* PERT chart scheduling)
program logic, 61, **62**
Project Managers (PM), 1
project scheduler's role, 1-11
  chain of command considerations, 10-11
  charting and scheduling, 9
  Construction Managers (CM), 1
  consultative capacity of project schedulers, 6
  evolution and development of project scheduling, 2-3
  integrated CPM software, 8, 9-10
  interlinking or "dovetailing" various activities, 10
  job logic, 7
  phases of scheduling, 7-8
  scheduling fundamentals, 6-11
  standard operating procedures (SOP), 3-4
  time manipulation tools, 4
  traits for success, 263-265
proposal scheduling, 20-21
public works, 131-132
purchase orders, 45
  Contract/P.O. Summary, 246, **247**
  long-lead items, 149

Q

quality assurance reports (QAR), 101, 116
quantity surveys, 136-137
quick ratio, 283

R

range commands, 168
ranges, data table ranges, 180, 188-189
ratios, business ratios, 281-284
record keeping, 46, 276
records, database operations, 183-184
recycling schedules, 157-158
relative addressing, 173
remote-access computing, 286-287
rentention, 284-285
reporting, 44, 45
  cost tracking reports, 25, 26
  daily field reports, 160-162, 227, **228**
  software for reporting, 23
  sorts, 164-166

residential construction example of CPM, 37, **38**, **39**
resource allocation, 59
results area, database operations, 182-183
retained earnings, 279
retention, 278
row height setting, 254

S

S-charts, 9, 25, 79-80, **81**, 138, 151
safety, 125, 129-131
saving your work, 178-179, 181-182, 248-249, 260-261
Schedule of Anticipated Earnings, 219, **220-222**
schedule planning, 155
schedule-operations analysis , 159-160
scheduling fundamentals, 6-11
scheduling systems, 13-23
  bar chart scheduling, 13-16, **15**
  benchmarking, 13
  computer requirements, 21-23
  field scheduling, 19
  integrated CPM systems, 22-23
  local area networks (LAN), 22-23
  logic-diagram chart scheduling, 13
  macroactivities vs. microgrouping, 14, 16
  milestones, 13, 22
  network requirements, 22-23
  PERT chart scheduling, 13, 16-17
  proposal scheduling, 20-21
  scheduling philosophy, 18
  selection of scheduling system, 21-22
  training requirements, 21-22
scientific notation numbers, 171
selection of scheduling system, 21-22
Shop Drawings Log, 236, **237**
shop drawings, blueprints, 147-149
  logbook of drawings, 154
  Shop Drawings Log, 236, **237**
shortcut commands, 179
sick leave, 127, 300
site scheduling (*see* field scheduling)
slack time (*see* float)
Small Business Administration (SBA) loans, 285-286
Social Security Numbers, 129, 302
sole ownership of business, 273
Sort by Activities, 80, 190, **191-194**
Sort by Early Starts, 223, **224-226**
Sort by Events, 195, **196-199**
Sort by I-J Numbers, 200, **201-204**
Sort by Job Logic, 205, **206-209**
Sort by Total Float/Late Start, 75, **76**, 210, **211-214**
sorting databases, 164-166, 186-187
  closing a sort, 249
  file management, maintaining schedule sorts, 258-260
  opening a sort, 250
  saving sorts to a different disk drive, 260-261
  saving sorts to different subdirectory, 260
  saving sorts, 248-249
specifications and standards, 6, 144, 146
  EJCDC table of contract standards, 115, 331-343
speed/shortcut commands, 179
splitting cells, 252-253
spreadsheet data into databases, 185
standard operating procedures (SOP), 3-4
start dates, 31, 44, 65, 67, 75
  Bar Chart by Early Start, 229, **230-233**
  Sort by Early Starts, 223, **224-226**
  Sort by Total Float/Late Start, 75, **76**, 210, **211-214**
stock, 279
stockholder's equity, 279
stop notices, 123-124, 132
strikes and work stoppages, 126, 127, 128, 300, 301
string operators, 175, **176**, 177
subcontractors, 3, 44, 59
  claims against subcontractors for delays, 99, 100-104
subdirectories, 260
Submittal Items Tracking, 238, **239**
subtasks, 63
supervisors, 59
suspension of project schedule, 296-297
syntax errors in data entry, 169-170

T

takeoffs, 135-136
tasks, 30, 63
taxes, 265-268, 274
  deductions for expenses, 268
  home office deduction, 267
  withholding taxes, 127, 300
time management, 4, 33, 44
time-scaled charts, 32
timeline computations, 28, 29, 35, 37, 65, 67, 155
total float, 31, 70, 72, 75
  Sort by Total Float/Late Start, 210, **211-214**
total liabilities to total assets ratio, 284
training in CPM, 21-22
transmittals, 154-155
  Correspondence Transmittals, 240, **241**
trends analysis, 139-140
  Schedule of Anticipated Earnings, 219, **220-222**

U

Undo command, 181-182
unemployment insurance payments, 129, 302-303
unforeseen (changed) conditions, 292
union vs. non-union labor, 125
updating schedules, 44

V

vacation pay, 127, 300
values, 168
variables, database operations, 182
velocity diagrams, 9, 29, 32, 77-89
  activity-on-node diagramming, 87-88, **89**
  arrow diagram vs. precedence diagram, 83, **84**, 87
  bar charts, 79, 82
  constraints, 77
  dependency bar chart scheduling, 88
  fast-tracking, 78
  flowlines, 79-80, **81**
  I-J numbers, 78, 83, **84**, 85, **86**, 88, **89**
  job logic, 77
  multiple critical paths, 78
  Network Timeline, 234, **235**
  phase scheduling, 77
  precedence diagramming, 83, 85, **86**, 87
  S-charts, 79-80, **81**
  Sort by Activities timeline, 80
vested parties, 10

W

wages, payment of wages due, 126-127, 299, 300
what-if analysis, 189
Windows use of FastPro (*see also* FastPro), 255-258
withholding taxes, 127, 300
WYSIWYG format, 168

About the Author

Jonathan F. Hutchings, BCM, is CEO of Construction Management Institute, Santa Cruz, California. He is a licensed building contractor and holds a degree in construction management. He holds state teaching and administrative credientials in construction management, business management, and microcomputer applications. He teaches construction law, contract law, and labor codes at Contractors State License Schools, San Jose campus. He also teaches construction management for the U.S. Department of Labor.

Installation Instructions

This software program is configured to run on the PC spreadsheet programs Lotus, Excel, or Quattro (versions 3.0 or higher), on a Windows platform (version 3.1 or higher). Boot your computer, open your spreadsheet program, then insert the disk and call up the files you want to use.

To reduce access time, you can also copy files from the disk to your hard drive through standard Windows copying procedures. (In Windows, go to the File Manager and call up the contents of your disk. By holding down the Ctrl key and clicking, you can define more than one file at once. By holding down the button and dragging the outline, you can pull the disk's contents to another drive or subdirectory.) Copying your disks to the hard drive will produce greater speed in saving and opening the files, and in compiling the summary sorts as well.

The disk contains the following files:

1STEP.WK4
CPMSORT1.WK4
CPMSORT2.WK4
CPMSORT3.WK4
CPMSORT4.WK4
CPMSORT5.WK4
CPMSORT6.WK4
CPMSORT7.WK4
CPMSORT8.WK4
CPMSORT9.WK4
SORT10.WK4
SORT11.WK4
SORT12.WK4
SORT13.WK4
SORT14.WK4
SORT15.WK4
SORT16.WK4
SORT17.WK4

DISK WARRANTY

This software is protected by both United States copyright law and international copyright treaty provision. You must treat this software just like a book, except that you may copy it into a computer in order to be used and you may make archival copies of the software for the sole purpose of backing up our software and protecting your investment from loss.

By saying "just like a book," McGraw-Hill means, for example, that this software may be used by any number of people and may be freely moved from one computer location to another, so long as there is no possibility of its being used at one location or on one computer while it also is being used at another. Just as a book cannot be read by two different people in two different places at the same time, neither can the software be used by two different people in two different places at the same time (unless, of course, McGraw-Hill's copyright is being violated).

LIMITED WARRANTY

Windcrest/McGraw-Hill takes great care to provide you with top-quality software, thoroughly checked to prevent virus infections. McGraw-Hill warrants the physical diskette(s) contained herein to be free of defects in materials and workmanship for a period of sixty days from the purchase date. If McGraw-Hill receives written notification within the warranty period of defects in materials or workmanship, and such notification is determined by McGraw-Hill to be correct, McGraw-Hill will replace the defective diskette(s). Send requests to:

McGraw-Hill, Inc.
Customer Services
P.O. Box 545
Blacklick, OH 43004-0545

The entire and exclusive liability and remedy for breach of this Limited Warranty shall be limited to replacement of defective diskette(s) and shall not include or extend to any claim for or right to cover any other damages, including but not limited to, loss of profit, data, or use of the software, or special, incidental, or consequential damages or other similar claims, even if McGraw-Hill has been specifically advised of the possibility of such damages. In no event will McGraw-Hill's liability for any damages to you or any other person ever exceed the lower of suggested list price or actual price paid for the license to use the software, regardless of any form of the claim.

McGRAW-HILL, INC. SPECIFICALLY DISCLAIMS ALL OTHER WARRANTIES, EXPRESS OR IMPLIED, INCLUDING, BUT NOT LIMITED TO, ANY IMPLIED WARRANTY OF MERCHANTABILITY OR FITNESS FOR A PARTICULAR PURPOSE.

Specifically, McGraw-Hill makes no representation or warranty that the software is fit for any particular purpose and any implied warranty of merchantability is limited to the sixty-day duration of the Limited Warranty covering the physical diskette(s) only (and not the software) and is otherwise expressly and specifically disclaimed.

This limited warranty gives you specific legal rights; you may have others which may vary from state to state. Some states do not allow the exclusion of incidental or consequential damages, or the limitation on how long an implied warranty lasts, so some of the above may not apply to you.